Building Genetic Medicine

Inside Technology

edited by Wiebe E. Bijker, W. Bernard Carlson, and Trevor Pinch

A list of the series appears on page 263.

Building Genetic Medicine

Breast Cancer, Technology, and the Comparative Politics of Health Care

Shobita Parthasarathy

The MIT Press
Cambridge, Massachusetts
London, England

For information on quantity discounts, email special_sales@mitpress.mit.edu.

Set in Stone serif and Stone sans by SNP Best-set Typesetter Ltd., Hong Kong. Printed and bound in the United States of America.

Library of Congress Cataloging-in-Publication Data
Parthasarathy, Shobita.
  Building genetic medicine : technology, breast cancer, and the comparative politics of health care / Shobita Parthasarathy.
    p. ; cm.—(Inside techology)
  Includes bibliographyical references and index.
  ISBN-13: 978-0-262-16242-5 (hardcover : alk. paper)
  1. BRCA genes—Diagnostic use—United States. 2. BRCA genes—Diagnostic use—Great Britain. 3. Genetic screening—Social aspects—United States.
4. Genetic screaning—Social aspects—Great Britain. 5. Breast—Cancer—Diagnosis—Social aspects—United States. 6. Breast—Cancer—Diagnosis—Social aspects—Great Britain. I. Title. II. Series.
  [DNLM: 1. Genetic Screening—Great Britain. 2. Genetic Screening—United States. 3. Breast Neoplasms—prevention & control—Great Britain. 4. Breast Neoplasms—prevention & control—United States. 5. Cross-Cultural Comparison—Great Britain. 6. Cross-Cultural Comparison—United States. 7. Delivery of Health Care—Great Britain. 8. Delivery of Health Care—United States. 9. Ovarian Neoplasms—prevention & control—Great Britain. 10. Ovarian Neoplasms—prevention & control—United States. QZ 50 P273b 2007]
  RC268.44.B73P37   2007
  362.196′04207—dc22

                                                                2006027956

10 9 8 7 6 5 4 3 2 1

Contents

Acknowledgments

I have always felt that I was very lucky to find this story, and the truth is that I would never have been able to explore or understand it properly without the support of a dense network of individuals, organizations, and institutions. First and foremost, I must thank the scientists, physicians, genetic counselors, activists, industry representatives, academics, and government officials who agreed to give up their valuable time to speak with me about the development of genetic testing for breast cancer. Meeting with them, sometimes over many hours and multiple appointments, provided me with a rich and nuanced perspective which I would have never gotten from published documents or media accounts that were more easily available. Some even opened up their personal archives to me, helping me find rare but pivotal documents that allowed me to piece together the history and politics of this new technology. They literally made the story come alive, and helped me develop a sensitivity to the issues that drove their actions.

I am also grateful to the people who enriched my research and analysis simply through everyday conversation; I was fortunate that many felt they could share their personal anecdotes and emotions with me outside the interview setting. Indeed, as I began work on this project, I quickly realized that everyone had an experience or opinion that was relevant to my analysis. Many told me about friends and relatives who had suffered from breast cancer and how concerns about whether they would pass the disease on to their families had generated guilt and anxiety. Occasionally, I was even put in the awkward role of genetic counselor, as people worried aloud about the consequences of a family history of disease for their own health or what might happen if information about their genomes fell into the wrong hands. I was asked many pointed and insightful questions—some of which I could answer, some I could not, but all of which motivated me to ask more questions, do more digging, and think harder about what it all meant. They reminded me of the realities of the disease experience, the high stakes of biomedical research and technological development, and the issues that the languages of DNA and molecular genetics are raising—and that we are only beginning to address.

Without the help of my dissertation committee, I would have never known how to develop this project, interpret the meaning of my research findings, or articulate my conclusions. Stephen Hilgartner guided the project, helped me to develop effective strategies for both research and writing, and suffered through many drafts. He exhibited tremendous patience with me, and taught me to develop a critical perspective toward my own work. Sheila Jasanoff was and continues to be a great inspiration to me. She held me to exacting standards but provided me with the example and encouragement to reach them. I am also grateful to Sidney Tarrow, who helped me think about the meaning and implications of my work beyond the science and technology studies community and challenged me to explain myself to a broad audience. Peter Dear went far beyond his duties to advise me on crafting my theoretical approach and language carefully.

I have also benefited greatly from intellectual exchanges with colleagues in both the United States and Europe, who helped me think through analytical and empirical dilemmas, asked me difficult questions, and commented on parts or all of the book manuscript. Thanks go to Pascale Bourret, Michael Dennis, Robert Doubleday, Paul Erickson, Sally Gibbons, Tom Gieryn, Herbert Gottweis, David Guston, Rob Hagendijk, Neil Holtzman, Joel Howell, Pauline Kusiak, Javier Lezaun, Alicia Löffler, Marybeth Long Martello, Michael Lynch, Anna Maerker, Sofia Merajver, Jon Merz, Clark Miller, Nelly Oudshoorn, Miranda Paton, Dan Plafcan, Judith Reppy, Dan Sarewitz, Jessie Saul, Sonja Schmid, Sara Shostak, Lucy Suchman, Sandy Sufian, John Tresch, David Winickoff, and Norton Wise. Many of them were simultaneously friends and colleagues, and helped keep me going when I couldn't see the light at the end of the tunnel. At The MIT Press, I would like to thank Trevor Pinch, Sara Meirowitz, and Paul Bethge, who confidently shepherded a first-time author through the publishing process.

The manuscript has evolved through numerous presentations at seminars and conferences organized by the American Sociological Association, Columbia University, Cornell University, Duke University, the European

Association for the Study of Science and Technology, the London School of Economics, the American Association for the Advancement of Science, the National Institute of Environmental Health Sciences, Northwestern University, the Science and Democracy Network, the Society for the Social Studies of Science, the Society for the Study of Social Problems, the Society for the Study of Symbolic Interaction, the Swiss Association for the Studies of Science, Technology, and Society, the University of California at Los Angeles, the University of Cambridge, the University of Chicago, the University of Illinois at Chicago, the University of Missouri at Columbia, the Virginia Polytechnic Institute, Wells College, and York University (Toronto). I thank them all for the opportunity to present and discuss my work-in-progress.

I would not have been able to conduct the research or writing for this manuscript without financial and institutional support from multiple sources. While at Cornell University I received funding from the Department of Science and Technology Studies, the Einaudi Center for International Studies, the Ethical, Social, and Legal Implications Program of the Genomics Initiative, the Institute for European Studies, the Women's Studies Program, and the Graduate School. I also received a Dissertation Improvement grant from the National Science Foundation (SES-9906183) to conduct my fieldwork. The Kennedy School of Government at Harvard University and the Centre for Family Research at the University of Cambridge granted me workspace and institutional affiliation as I conducted my dissertation research. I completed the writing stage of my dissertation at Duke University, as part of the New Beginnings Program in Science, Culture, and Society and the Department of Economics.

Postdoctoral fellowships from the Science in Human Culture Program at Northwestern University, the Wellcome Trust's Biomedical Ethics Programme (GR068437MA), and the Center for Society and Genetics at the University of California at Los Angeles helped me continue my fieldwork and turn the dissertation into a book. While writing, I received much-needed office space from the Department of Sociology and Center for Business, Government, and Society at Northwestern University and the Centre for Family Research at University of Cambridge. I completed the manuscript and underwent the review and revision process while at the Gerald R. Ford School of Public Policy at the University of Michigan.

I am particularly grateful to those who made this obviously nomadic journey a pleasant and minimally disorienting one. Martin Richards at

Cambridge was a wonderful mentor to me during my many trips to Britain, and provided me with a valuable perspective when he read and commented on my manuscript. Ken Alder was a great friend and colleague during my time in the Science in Human Culture Program at Northwestern University. Also at Northwestern, Daniel Diermeier at the Center for Business, Government, and Society took a chance on a junior colleague and taught me how to make my ideas and expertise matter to the business world. I also owe special thanks to Lillian Isacks, Deb Van Galder, and Judy Yonkin at Cornell, Peter Sedlak at Harvard, Jill Brown at Cambridge, Elena Glasberg and Caroline Light at Duke, Kristen Leuking and Phyllis Siegel at Northwestern, Carlene Brown and Greta Nelson at UCLA, and Sharon Disney and Andrea Perkins at University of Michigan. Colleen Betsy Popken and Monamie Bhada, my fantastic research assistants, provided invaluable help.

Finally, I must thank the friends and family who kept me both working and laughing, helping me to focus on my work and distracting me when necessary. Jamey Wetmore and Carla Bittel have been my partners in crime for over ten years now, first riding the roller coaster of graduate school with me, and continuously reminding me that academia need not be an isolated or lonely place. Academic compatriots, commiserators, and fellow adventurers, they have supported me in work, in life, and the majority of time I spend in between. Mona Jhawar deserves special recognition, not only because she was always just a phone call away, but also because she read parts of the manuscript that all-important last time. Carlos Cortez, Ruth Farrell, Mary Nam, Nithya Ramanathan, Lucy Stanfield, and Arthi Varma were also wonderful friends throughout the book writing process. My brother played an important role, always making sure that I didn't take myself too seriously and escaping with me whenever I asked—sometimes even to exotic locales. I dedicate this book to my parents, who told me that I can accomplish anything I set my mind to and have shown me, through their examples, how to reach my goals. My father taught me about drive, tenacity, hard work, and integrity, and remains a valuable advisor and confidant as I engage in the challenging task of growing up. My mother kept me motivated, never let me give up, and shared my joys and frustrations as she became a book author herself. You both gave me so much more than life, you opened my eyes.

Building Genetic Medicine

Introduction: A Framework for the Comparative Analysis of Technology

National context matters in the development and use of science and technology. Although knowledge and innovation are often thought to provide the foundations for the processes that many refer to as globalization—consider, for example, the World Wide Web, the international space station, transnational airplane travel, and even the multinational pharmaceutical industry—countries have adopted very different approaches to issues such as stem-cell research and euthanasia.[1] In fact, different interpretations of the safety of genetically modified organisms and the consequences of climate change, among other things, have led to vigorous transnational conflicts that have not only scientific but also social, political, and economic consequences.[2] But exactly how does national context matter? In what ways do these national approaches influence how research is conducted and technologies built? Why do national borders still matter, even as countries are becoming more closely knitted together? What are the consequences of national approaches to science and technology for globalization, and for our daily lives?

In this book, I explore these questions through a comparative analysis of one of the most highly anticipated and publicized technologies of late-twentieth-century medicine: genetic testing for breast and ovarian cancer. Human genetics, in particular, is often characterized as the ultimate global endeavor. Transnational scientific investigations into the human genome have characterized DNA as made up of universal building blocks that lead to genetic similarities among humans—we are thought to share 99.9 percent of our genetic makeup. Furthermore, many have suggested that our common heritage provides a blueprint for curing disease around the world. In June 2000, in an announcement celebrating the completion of the first draft DNA sequence of the human genome, US President Bill

Clinton spoke in such terms: "Without a doubt, this is the most important, most wondrous map ever produced by humankind. . . . With this profound new knowledge, humankind is on the verge of gaining immense, new power to heal."[3] The global implications of genomics have also been considered by the United Nations, which commissioned a report on Genomics and Global Health as part of its Millennium Development Program. In 2003, UN Secretary General Kofi Annan stated: "Recent advances in . . . genetics and biotechnology hold extraordinary prospects for individual well-being and that of humankind as a whole."[4] By its objective and universally relevant nature, exploration of the human genome seems to be a quintessential example of the importance of science and technology in globalization, building bridges across countries and promoting common interests.

Despite the global nature of both scientific endeavors and potential health implications in this field, however, scholars have demonstrated that policies related to genomics and genetic medicine are shaped by national context. The anthropologist Paul Rabinow, for example, has described how in the 1990s the French government chose to dismantle a transatlantic public-private research partnership devoted to finding genes linked to diabetes because of national ideas about the role of benevolence in medicine and the commodification of the human body.[5] In Iceland, a private company's efforts to create a data bank for genetic research that was made up of citizens' DNA led not only to a struggle shaped by national political and media institutions, but also to a public debate that referred often to a unique Icelandic identity, history, and culture.[6]

I argue that the influence of national context is felt far beyond public policy and political debate to the level of practice, fundamentally influencing human genome science and technology. Through a comparison of how genetic testing for breast and ovarian cancer (known as BRCA testing) was built in the United States and in Britain, I develop three arguments. First, I demonstrate that national context plays an important role in the development of science and technology, not merely in terms of its regulation but also in terms of how practices and artifacts are shaped. Second, I complicate most predictions of our genetic future by showing that genetic medicine is being built quite differently according to national context and that these variations have important consequences for our lives and for health care. In particular, I show that these national differences in how breast cancer genetic science was conducted and understood and how BRCA-testing technologies were built have influenced how genetic medicine is organized and regulated, how users are envisioned, and how risks and disease are being defined and redefined. Finally, I argue that these deeply embedded national differences in science and technology can help to explain some of the challenges to transnational technology transfer that are beginning to occur around the world in domains such as trade, intellectual property, and drug safety.

Finding the Breast Cancer Genes

Breast cancer is the most frequently diagnosed cancer worldwide, and is heavily publicized as the leading cause of cancer death for women in Britain and the second leading cause of cancer death for women in the United States.[7] The news media, medical charities, and public health organizations in each country quote that a British woman has a 1-in-12 chance of developing invasive breast cancer during her lifetime, and that an American woman has a 1-in-8 chance. Although governments and medical charities (including the Imperial Cancer Research Fund and Cancer Research Campaign in Britain and the National Cancer Institute and American Cancer Society) have spent a considerable amount of money to look for a cause and to develop prevention and treatment strategies for the disease, neither an unequivocal cause nor a completely effective prevention, detection, or treatment strategy has yet been found. Mastectomy, which involves breast removal, not only has severe physiological and psychological effects but also doesn't completely eliminate the incidence of a first breast cancer or even the risk of recurrence.[8] And radiation therapy and chemotherapy (a chemical treatment to kill tumor cells)—uncomfortable procedures that induce considerable sickness and hair loss among patients—are still not completely reliable.[9] Even technologies to detect cancer early are not entirely reliable. Mammographic screening, which involves breast imaging using low levels of radiation to detect tumors, is controversial; some argue that it is not useful to identify tumors in the denser breasts of young women, while others suggest that it does not reduce the number of cancer deaths and that therefore it should not be considered an answer to the disease.[10]

Within this environment of imperfect prevention, detection, and treatment measures, vigorous advocacy in support of the needs of breast cancer patients has emerged. Women began to articulate their discontent with treatments for breast cancer in the early 1970s. As feminists and women's health activists encouraged women to take control of their health care, prominent breast cancer patients, including the feminist Betty Friedan, the journalist Rose Kushner, and the writer Audre Lorde, questioned the need for and the efficacy of the Halsted mastectomy, a radical procedure that had been the primary treatment for the disease for decades and which entailed removal of not only breast tissue but also muscle and part of the chest wall.[11] In the 1980s and the 1990s, breast cancer patients, observing the successes of the well-organized and powerful AIDS activist movement, began to mobilize in large numbers, this time with far-reaching social and political goals: to increase public and government attention to their disease as an epidemic that was affecting not only women but also their families and friends. The San Francisco-based organization Breast Cancer Action (BCA), founded in 1990, quickly became quite influential. In 1991, one of BCA's founders, Elenore Pred, became the first breast cancer activist to address the President's Cancer Panel. In 1993, BCA helped draft and enact California's Breast Cancer Act, which raised money for screening and research and guaranteed that advocates would participate in research funding decisions. In 1991, the National Breast Cancer Coalition, a nationwide network made up of survivors, physicians, support groups, and charities and based in Washington, was formed. Its advocacy efforts began in earnest with a letter-writing campaign that led the US Department of Defense to create a $210 million research program devoted to breast cancer research.[12] Largely as a result of the NBCC's activities, the US government's funding for breast cancer research increased from approximately $90 million in fiscal year 1990 to more than $800 million by fiscal year 2003. Since then, breast cancer activists have organized a variety of campaigns to increase awareness of the disease, and their efforts have been successful: breast cancer is now the most commonly discussed disease among women.[13] It should not be at all surprising, then, that research to find genes linked to breast cancer was highly anticipated and publicized; the genes were simultaneously of scientific, medical, and public interest.

Starting in the 1980s, scientists from the United States, England, France, Germany, Japan, and other countries participated in what was often referred to as an international "race"[14] to discover and identify the nucleotide sequences linked to breast cancer, perhaps anticipating that the winner of the race might enjoy both professional and financial rewards. By the early 1990s, discovery of the gene seemed imminent when a group led by Mary-Claire King at the University of California at Berkeley found that Breast Cancer Susceptibility Gene 1 (BRCA1), the first gene linked to breast cancer, lay somewhere on chromosome 17.[15] With King's announcement, investigations intensified. Transnational collaborations formed and disintegrated. In September 1994, a group led by scientists at Myriad Genetics (a biotechnology company based in Salt Lake City) and including researchers from the National Institute for Environmental Health Sciences, a subdivision of the National Institutes of Health, announced the mapping and the sequencing of the BRCA1 gene, which seemed to be linked to inherited breast and ovarian cancer.[16]

The news was met with considerable excitement. The American television network NBC deemed the story so important that it reported the news on September 13, days before the journal article based on the discovery had even completed the formal review process at *Science* magazine. Tom Brokaw opened his newscast that night by saying: "There's an important breakthrough in breast cancer research. . . . A rogue gene could show the way to treatment and prevention." Within hours, other television networks had reported the story, and soon it began to appear on the front pages of newspapers and magazines around the world.[17] The *New York Times*, for example, trumpeted: "Capturing a genetic trophy so ferociously coveted and loudly heralded that it had taken on a near-mythic aura, a collaborative team of researchers has announced the discovery of a gene whose mutation causes hereditary breast cancer."[18] The discovery even inspired Harold Varmus, then director of the National Institutes of Health, to state, at a press conference: "This is an extremely important development in the understanding of breast cancer. . . ."[19] The extensive media coverage, coupled with Varmus's announcement, underscored how the discovery had been deemed to be of considerable historical, social, and scientific significance.

Notwithstanding the enthusiasm over the BRCA1 gene discovery, many scientists believed that there was at least one more gene linked to breast cancer, and continued their research. In December 1995, investigators led by a group at Britain's Institute for Cancer Research announced that they had mapped and sequenced the BRCA2 gene, which was linked to incidence of ovarian cancer and female and male breast cancer.[20] The media announced this second discovery with almost as much excitement as the first. An article in the *Financial Times* noted: "Scientists at the Institute of Cancer Research in London have won the most competitive race in medical research this year—to isolate the second gene responsible for inherited breast cancer."[21] Together, the discoveries of the BRCA1 and BRCA2 genes were seen to lay the foundation for a new era of medicine in which a woman's genetic makeup would guide the prevention and treatment options made available to her.

The BRCA gene discoveries were considered scientifically and medically important because they demonstrated that genes influenced the incidence of common diseases, not only that of rare disorders. Although only 5–10 percent of the individuals who contracted breast cancer did so because of a BRCA mutation, and some wondered if the breast cancers that affected those with BRCA mutations were of a different type (affecting younger people, for example), many hoped that understanding the mechanisms by which the BRCA genes worked would provide insight into all breast cancers. Even if someone had a BRCA gene mutation, however, her future cancer prognosis was not clear. Although research is ongoing, scientists believe that normal BRCA genes act as tumor suppressors, repairing cells that have been damaged through environmental or other means. When a BRCA gene is mutated, a mammary cell is left unable to repair its DNA and is therefore left vulnerable to an assault that could lead to uncontrollable cell growth and eventually a malignant breast tumor. Not only does disease incidence for BRCA mutation-positive individuals involve other, often random factors, but scientists have found hundreds of alterations to these long and complex BRCA genes and each does not necessarily debilitate the gene in the same way or even always leave a cell vulnerable to further DNA damage. It is difficult to discern, then, the exact relationships between individual mutations and risk of future disease. Although little such targeted information is available, studies to date suggest that the lifetime risk of breast cancer for individuals with a BRCA mutation can vary from 36 percent to 85 percent.[22]

Because the BRCA genes only provided information about disease risk (rather than certainty of future incidence), and because they were linked to a common disease with a high profile, they raised a number of questions about the use of genetic information for prevention or treatment when they were discovered in the mid 1990s. Would such information have any utility? What would the psychological implications be? Did effective preventive options need to be available to justify availability and use of genetic tests? How should genetic risks be balanced with the risks of medical interventions geared toward prevention? The development of BRCA testing would begin to provide answers to these questions, which would surely condition the continuing emergence of genetic medicine.

Understanding Genetic Medicine

The discoveries of the BRCA genes in the mid 1990s occurred amidst considerable scientific attention to and public interest in human genomics and genetic medicine. The Human Genome Project, a transnational effort to map and sequence the human genome, was well underway, and genes for a variety of conditions—among them heart disease, obesity, diabetes, and homosexuality—were being hunted. These large financial commitments to human genetics investigations by both the private and public sectors, however, were and continue to be accompanied by public and scholarly discussions trying to make sense of the social, ethical, political, and legal dimensions of this new area of science and technology and its implications for medicine and society.

First, while these observers share the excitement of scientists and doctors that human genetics research might lead to the diagnosis and treatment of disease risk long before its incidence, many also worry that innovations to identify and manage individuals at the level of their DNA will lead to the reconstruction of social orders along genetic lines, with individuals being defined medically, socially, and politically according to their genetic makeup. This phenomenon has been variously described as geneticization, genetic essentialism, and genetic exceptionalism.[23] The women's health scholar Abby Lippman writes:

Geneticization refers to an ongoing process by which differences between individuals are reduced to their DNA codes, with most disorders, behaviors and physiological variations defined, at least in part, as genetic in origin. It refers as well to the process by which interventions employing genetic technologies are adopted to manage problems of health. Through this process, human biology is incorrectly equated with human genetics, implying that the latter acts alone to make us each the organism she or he is.[24]

This idea rests on ongoing observations by medical sociologists that Western scientific medicine is leading us through a gradual process of "medicalization," and more recently "biomedicalization," in which such social conditions as homosexuality, obesity, menopause, and pre-menstrual syndrome have been brought into the domain of health care because of the increasing dominance of medical infrastructures in late-twentieth- and early-twenty-first-century Western societies.[25] Observers of the development of genomic medicine worry that increasing attention to our genes will take us down a similar path, with an emerging corporate, scientific, medical, and political infrastructure focused on the information contained in our DNA.

"Geneticization," many argue, can have serious social consequences. The sociologist Troy Duster, among others, has wondered whether the identification of individuals as at risk for particular diseases through genetic testing will lead to the formation of a "genetic underclass," a group of genetically at-risk individuals whose rights will be restricted because of their propensity to disease.[26] At its extreme, the creation of a genetic underclass could lead to a reproduction of early-twentieth-century eugenics movements, or individuals may simply be turned away from employment opportunities or be denied health or life insurance because of their future prospects of disease. The passage of federal and state laws in the United States to protect individuals from genetic discrimination in employment and insurance and the creation of oversight committees to address these issues in Britain have not managed to calm these fears.

Second, many worry about the quality of health care that will accompany the development of genetic testing technologies.[27] As more genes are found for common diseases, and as more genetic tests become available, there will probably be a large population of individuals interested in using these technologies. How will health-care professionals and laboratories deal with the extraordinary demand likely to be generated by these discoveries? Will they explain to individuals that genetic testing, even for common diseases, is likely to be useful for only a very small number of people? How will health-care professionals explain the complicated information about risk and inherited susceptibility to individuals interested in testing? Will clinical care be restricted to geneticists who have specialized training, or will primary-care physicians be allowed to contend with the anticipated demand?

Third, how will test developers, health-care professionals, and regulators deal with the "therapeutic gap"—the paucity of interventions available for individuals who test positive, or "at risk," for genetic mutations for many common diseases?[28] The therapeutic gap was a particularly important issue in the case of breast cancer, where few proven preventive interventions seemed to exist.

Fourth, commentators wonder how these new technologies will be regulated. Genetics services for rare disorders have been offered on such a small scale that they have largely escaped government intervention. How will genetic medicine for common conditions fare? What logic will policymakers use as they develop regulations in this area? Will they understand genetic testing as similar to existing clinical services, medical devices, or drugs? Will government regulators limit access to genetic testing? On what basis? Fifth, as information about people's genomic makeup is generated, who will control access to this information? Do participants in human genetic research have ownership rights in their DNA? What ownership rights do researchers have? How far should ownership rights over genetic research and technology extend? Who will control access to the genetic information that is generated: physicians, insurers, government, or the individual?

The BRCA genes were found as such questions were being articulated and vigorously discussed around the world. Many saw the development of the first genetic test for a common disease that would be of interest to a large number of people as the perfect test case. It seemed to raise all these questions, and it seemed to have implications for a large population of people concerned about their or their loved ones' risks of contracting breast or ovarian cancer. As a result, development of the new technology quickly became a battleground over how to settle these questions not only in the case of breast cancer but also for the future of genetic medicine. An editorial published after the discovery of the first BRCA gene concluded:

When the first "disease genes" were identified and a few people had to decide if they wanted to know whether they were doomed by their DNA, ethicists would shake their heads and say, "If you think these are tough issues, just wait till they find the breast-cancer gene." When it was first clear that genetic tests could enable employers and insurers to screen for inherited, truly pre-existing, conditions, someone would say, "Just wait till they find the breast-cancer gene." When parents faced the dilemma of whether to bring children into the world as genetic heirs to rare illnesses, someone would say, "Just wait till they find the breast-cancer gene." Now the waiting is over. . . . Scientists on earlier voyages over the vast new sea of genetic knowledge had discovered small islands of illness like Huntington's and cystic fibrosis. But these geneticists have discovered the mainland: breast cancer. . . . As of today, the hopes, the traumas and the uncertainties raised by genetic knowledge are no longer limited to exotic diseases. They are becoming part of our everyday, garden variety lives.[29]

The way genetic testing for breast cancer was built would likely have serious and significant implications, not only for those concerned about their BRCA risk, but also for everyone who might use genetic medicine in the future.

Throughout this book, I explore the development of BRCA testing, investigating how the issues outlined above were debated and resolved, and how these resolutions were reflected in the way the new technology was built and regulated. By approaching this case study with a cross-national lens, we can also see how the debates and technological development compare in the two countries. Is geneticization an emerging global phenomenon that is being produced in tandem with genetic medicine? Or is it specific to national contexts? Are the same types of questions and criticisms of genetic medicine being raised in different countries, and what are the implications for the way genetic technologies are built? Through this in-depth investigation, we will begin to see not only how the future of genetic medicine is beginning to take shape and compare across countries, but also how national contexts influence scientific and technological development.

A Framework for the Comparative Analysis of Technology

In order to compare genetic medicine in the United States and Britain, we must first develop a framework for this analysis. The idea that innovation is shaped by national context has long been discussed by scholars in political science and in economics. Many observers of comparative politics, for example, have argued that countries have distinct national styles of regulation—including institutional arrangements and decision-making approaches of the State—which lead to national differences in science and technology policy. In his comparative analysis of environmental policy in the United States and Britain, for example, David Vogel argues that the two countries adopted very different regulatory "styles." "British regulation," he notes, "is relatively informal and flexible while American regulation tends to be more formal and rule-oriented."[30] Ronald Brickman and his colleagues have echoed Vogel's observations in their investigations of chemicals regulation in Europe and the United States.[31] Sheila Jasanoff has also suggested that national differences lead to distinct ways of resolving scientific uncertainty for policy-making purposes. US regulators, for example, appeal to formal analytic and quantitative methods, while their British counterparts are more accepting of qualitative evidence and the subjective judgments of experts.[32] These differences often have important consequences, leading not only to policy differences but even to differences in how a problem is "imagined, characterized, delimited, and controlled."[33] In economics, scholars who follow the national-innovation-system approach suggest that the roles and activities of national institutions, both public and private, shape a country's propensity and capacity to innovate. Patel and Pavitt, for example, describe such a system as "the national institutions, their incentive structures and their competencies, that determine the rate and direction of technological learning (or the volume and composition of change-generating activities) in a country."[34]

These approaches, however, have two shortcomings that are particularly problematic for the questions this book seeks to answer. First, they privilege the roles of institutions, both public and private, in the innovation process. Second, they focus on decisions to encourage or regulate development of science and technology, but do not explore whether there are important differences in the practices or artifacts themselves—in the way research is done, or in the way technologies are built.

Beyond Institutions

Although institutions, both public and private, clearly play a pivotal role in the development of science and technology, it is also important to consider how culture, in the forms of shared understandings, traditions, and histories, provides meaning to and helps to sustain those institutional structures on which so many national comparative analyses hang. In a recent book comparing the development of biotechnology in the United States and Europe, Sheila Jasanoff notes that social entities such as "the state" must be seen as "historically situated, contingent, dynamic constructs, whose form and fixity are as much in need of explanation as they are available for explaining other developments."[35] Jasanoff argues for an understanding of national context in terms of political culture, moving beyond institutions to encompass "institutionally sanctioned modes of action, such as litigiousness in the United States, but also the myriad unwritten codes and practices with which a polity supplements its formal methods of assuring accountability and legitimacy in political decision-making."[36] The meanings of patent laws, for example, are contained not only in written words, but also in the way they have come to be understood and used over time in the context of industrial histories and traditions of commercializing research.

By taking into account both the structural and the cultural elements that figure in the development of genetic medicine, we will see a much more comprehensive and nuanced picture. We can go beyond regulatory frameworks to explore, for example, how particular traditions of providing health care influenced both policies and the development of a new medical technology. We can also investigate how histories of patient activism shaped not only the involvement of contemporary advocates, but also the actions of innovators, regulators, and health-care professionals. We can really explore technological development from the bottom up, rather than through the windows of institutions high off the ground.

Artifacts and Practices

As was described above, comparative analyses of science and technology traditionally focus on differences in regulatory policies or the amount of innovative activity that takes place in a country. They have not yet penetrated the "black box" of science and technology to determine whether its contents have been changed or assembled differently because of what lies outside. Given what we know about science, technology, and health policy, we might expect that national context would shape the regulation of a technology or an individual's access to it, but how does it influence the way technologies are built?

Scholars of science and technology studies have long argued that the substance of technological artifacts is shaped by social action, and a number of approaches have been proposed to help us investigate the relationship between technology and society. The Social Construction of Technology (SCOT) approach argues that "relevant social groups" play an important part in the development of technologies.[37] Rather than thinking about technologies as the result of a linear, pre-determined developmental path, SCOT asks us to consider the variety of groups who deem themselves important to the construction of a particular technology and their articulations of how it should be built, thus emphasizing a technology's multidirectional potential. The strategies and influence of these groups then play an important role in how technologies are eventually developed, including the components that are used and how they are put together. Actor-network theory (ANT) also emphasizes the multidirectional potential of technologies, envisioning a heterogeneous network made up of human actors (e.g., patient advocacy groups) and non-human actants (e.g., a gene-sequencing machine) that must be assembled to form a successful sociotechnical system.[38] One particularly powerful actor or actant might force a specific configuration of the network and therefore a particular technological design. The "social worlds" perspective suggests that we explore how different groups of actors and actants with shared commitments offer multiple approaches to the development of a technology, and that we assess the relative power of these worlds by investigating how these approaches influence the eventual development of a technology.[39]

Although each of these approaches provides us with tools to look inside the black box, follow the development of a technology around, and understand the importance of a variety of social actors (e.g., individuals, institutions, and organizations) in the construction of technologies, they provide us with little guidance to understand how social and political environments influence these processes. Are there more stable elements of societies—for example, particular political traditions or histories of health care—that lie outside the creation of actor networks, or beneath the definition of relevant social groups or social worlds? Are the strategies developed by social actors based on others that have historically been successes or failures? How can and should we understand national differences in the development of technologies?

Historians and sociologists of technology have tried to develop ways to incorporate the broader social and political terrain into their analyses. Wiebe Bijker, a proponent of SCOT, has introduced the concept of "technological frames," directing our attention beyond interactions among actors to "the goals, the ideas, and the tools needed for action."[40] Whereas the technological frame can incorporate more diffuse social elements because it is external to the actors involved and the interactions among them, it still exists only in relation to the development of a specific technology. "A technological frame," Bijker notes, "does not reside in individuals—it is largely external to any individual, yet located at the level of a relevant social group. Thus a technological frame needs to be sustained continuously by actions and interactions. They are not fixed entities, but are built up as part of the stabilization process of an artifact. The building up of a technological frame mirrors the social construction of an exemplary artifact, just as much as it reflects the forming of a relevant social group."[41] The technological frame provides us with historical perspective for understanding, for example, the concurrent evolution of the bicycle and the social groups who shaped it, but it does not help us to identify the persistent elements of national social and political cultures that might be important to a comparative analysis of technology.

Thomas Hughes, the primary developer of the systems approach to technology, has also tried to address these issues by introducing the concept of technological "styles," which include "entrepreneurial drive and decisions, economic principles, legislative constraints or supports, institutional structures, historical contingencies, and geographic factors, both human and natural."[42] While this concept allows Hughes to take national specificities and persistent social elements into account, it still does not provide much guidance for understanding the relationship between national context and technological development. Considering national context in terms of "styles" might help us understand *what kinds of* structural and cultural elements figure in the development of a new technology, but we still need to understand *how* these styles affect the way new technologies are built. Do these styles determine the course of technologies? Or are they simply available for developers to use or discard, depending on their interests and goals?

Ann Swidler's concept of a cultural "toolkit" can provide us with some guidance as to how the structural and cultural elements of a particular country might influence the development of a technology. Swidler proposes that we understand how culture influences action by thinking about it as a toolkit that provides the raw materials from which people construct a particular "strategy of action": "Strategies of action are cultural products; the symbolic experiences, mythic lore, and ritual practices of a group or society create moods and motivations, ways of organizing experience and evaluating reality, modes of regulating conduct, and ways of forming social bonds, which provide resources for constructing strategies of action."[43]

Conceptualized in this way, a national toolkit would include such cultural elements as national histories of patient activism, traditions of health-care provision, and acceptance of the commercialization of research as well as structural components such as laws and regulatory frameworks. Developers of new technologies, like all members of a polity, would have access to the national toolkit and choose which elements to incorporate into their technologies based on their priorities and goals. This toolkit, then, does not direct action or place countries on a pre-determined national technological path, but rather provides builders of technologies with a finite range of options from which to choose their "strategies."

As it follows how genetic testing for breast cancer is built, this book will investigate the elements contained in a national toolkit and how they are used to shape innovation. It is important to remember, however, that the items contained in these toolkits are quite diverse and sometimes contradictory, and they change often. As Swidler notes, "all real cultures contain diverse, often conflicting symbols, rituals, stories, and guides to action. . . . A realistic cultural theory should lead us to expect not passive 'cultural dopes' but rather the active, sometime skilled users of culture whom we actually observe."[44] As we well know, multiplicity and heterogeneity are common inside national borders. In the United States, for example, both proponents and opponents of a national health-care system draw on deeply held but sometimes contradictory national ideals of independence and community.[45] We shall see throughout this book that participants in the development of genetic testing for breast cancer, even within each country, drew upon distinctly national but multiple, and sometimes even conflicting, resources as they envisioned the new technology.

While Swidler focuses on the use of the toolkit concept to understand approaches to such social phenomena as poverty, love, and religion, it might be most useful and informative for the study of science and technology. Because developments in science and technology often require societies to accept drastically new ways of thinking and doing, they constitute novel moments where social agents lay bare their toolkits and develop their strategies anew. Swidler characterizes these moments as periods of "unsettled lives." She writes: "In such periods, ideologies—explicit, articulated, highly organized meaning systems (both political and religious)—establish new styles or strategies of action. When people are learning new ways or organizing individual and collective action, practicing unfamiliar habits until they become familiar, then doctrine, symbol, and ritual directly shape action."[46] The development of genetic testing for breast cancer, as the first technology of its kind, constitutes a novel moment that exposes the types of elements within a national toolkit and the way they are used; in addition, the strategies of action constituted during this episode will surely have implications for a more "settled" future of genetic medicine.

Investigating the development of this new technology through the lens of national comparison will enhance our ability to understand how national toolkits figure in the development of science and technology. As we analyze the developments in the United States and Britain side by side, both the differences in the technologies and the reasons for these differences will come into sharp relief. These two countries offer particularly good sites for comparative analysis. They have many similarities, as both are English-speaking and affluent Western capitalist democracies with close ties to one another as well as many shared political traditions. Both countries have been vigorously involved in genetics and biotechnology research, demonstrated perhaps most clearly by their joint leadership in the Human Genome Project.[47] They have also been equally involved in research on the genetics of breast cancer, and they report similar incidences of breast and ovarian cancer in their populations. Despite these similarities, however, there are a few notable differences. The countries approach health care in distinct ways. The US has created a medical environment based primarily on private insurance, whereas in Britain most people use the government-run National Health Service. Also, although the US and Britain have similar patent laws, scientists in the two countries seem to have different approaches to the commercialization of research, with, for example, the US Patent and Trademark Office issuing 17 times as many biotechnology patents to American inventors as to their British counterparts.[48] To what extent do these similarities and differences matter? As this book progresses, we will discover not only the elements of each national toolkit but also how these toolkits figured in the development of the new technology.

What Is a Technology?

What exactly is genetic testing? Is it simply the methods used to analyze DNA in the laboratory? Does it also include the interaction between doctor and patient that takes place in the clinic and the computer programs that use statistical information to assess disease risks? Does it include only things (laboratory chemicals, machines, family-history questionnaires, consent forms)? Or does it also include people (patients, laboratory and clinic personnel) and places (specialized genetics clinics, general practitioners' offices)? This slipperiness in defining the object of this comparative analysis is not unique to genetic testing, and it has been discussed frequently by scholars of science and technology studies.[49] Donald MacKenzie and Judy Wajcman, for example, note that technologies have at least three layers of meaning.[50] Technologies can be physical artifacts, such as a computer or a vacuum cleaner. These technologies only have meaning, however, when they are part of a set of human activities. Thus, "technology" can include not only material artifacts but also the human actions that make sense of them. Finally, technologies can include what people know in addition to what they do, as the use of a technology also requires knowledge of *how* to use it, which may change according to individual needs and interests.[51] Thus, the boundaries of a technology can be defined in a number of different ways.

While there has been some conceptual traffic among the various approaches in the sociology of technology discussed above in the previous section, many of them draw different boundaries around the objects of their analysis. SCOT theorists usually focus on the production of material artifacts, researchers adopting the social worlds perspective investigate the production of both artifacts and meanings, and ANT and systems proponents conceptualize technologies as a network of both material artifacts and human actors. [52] These last two approaches do not make a priori conclusions about the relative importance of these human and non-human components. Rather, they challenge our conventional understandings of the roles of these elements, asking us to consider the functions of humans and non-humans in a similar manner. For the purposes of this study, I will define the technology of genetic testing for breast cancer to encompass the methods, materials, practices, and places involved in both the clinical and laboratory dimensions of the technology. This approach will accomplish two things. First, it will allow me to capture the multiple functions performed by a medical testing system. Medical testing systems involve so much more than the analysis of bodily materials in the laboratory. They must all somehow direct individuals to testing, assess their eligibility, inform them about potential risks, benefits, and implications of the test, extract material or information for testing by a technical apparatus, and report the results. Ultimately, based on the results, various medical management strategies may be implemented. These functions are carried out not only by material tools and techniques but also by people, places, and practices. Second, this broad definition will allow us to see and compare how people draw different boundaries and define BRCA-testing systems differently. Indeed, the way that medical testing systems carry out their functions can vary considerably depending on the specific test and its historical and geographical location. In the United States today, individuals might be directed to a medical test through the advice of a physician or by newspaper articles and direct marketing campaigns, while direct-to-consumer advertising is not likely to be part of the testing system in Britain where most people learn about new technologies through their physician or the media. Differences in how an individual learns about a technology might have important implications for how it is used—someone enticed to take a test by a company's marketing efforts might be much more likely to demand access and be convinced by the results of a test than someone advised about it by a physician. And one can easily envision other important differences. Insurance companies or the individual herself might assess whether a particular test is appropriate in the United States, while in Britain the National Health Service or a health-care professional might perform this task. The type of test available and how it is provided might also vary, depending on professional and regulatory histories; a test might be purchased through an over-the-counter kit, from a private stand-alone laboratory, or at a clinic. These differences might lead to different uses and meaning of the technology and also create different roles for physicians, test developers, and clients.

As these examples suggest, the functions of a testing system can be distributed over a diverse set of system components in many ways. The components chosen and the way they are fitted together to perform the functions of the medical testing system constitute what I will call a technology's *architecture*. Identifying the architectures of medical testing systems facilitates their comparison by highlighting how similar functions might include different assemblages of components or be carried out in different ways. Consider two common medical testing technologies:

amniocentesis and home blood-pressure testing. While equipment for testing blood pressure can be purchased in a drugstore and used by an individual in the privacy of her kitchen, amniocentesis must be conducted in a clinical setting and requires the participation of a laboratory to analyze the biological material. Examining the architecture of a technology, particularly in a comparative perspective, helps us to focus on similarities and differences in how the systems are constructed. This approach can also help us to understand exactly how social, political, and cultural contexts figure in their construction, and to focus on the consequences of each element of a technology's architecture.

We shall see throughout the book, for example, that the architectures of BRCA testing in both countries were built with a focus on offering women a test to analyze their genetic risk of contracting *breast* cancer. However, the BRCA1 gene is also thought to cause ovarian cancers, and the BRCA2 gene is also linked to ovarian and male breast cancer.[53] What accounts for this focus on the link between genetic testing and female breast cancer in both countries? First, the BRCA genes are thought to lead to many more female breast than ovarian or male breast cancers. This fact, coupled with the extensive public interest in breast cancer, has led most observers, informed or uninformed, to refer to the BRCA genes as "the breast cancer genes." Such an orientation has shaped political and biomedical attention to the development of BRCA testing. Breast cancer activists were much more involved than ovarian cancer activists, for example, in trying to influence development of the new test. Providers built their technologies in consideration of the high demand that would likely emerge for a test linked to breast cancer risk. Discussions about how to incorporate the test into medical care also focused on the implications for breast cancer prevention. Thus, in following the politics involved in building this new technology, I adopt this focus as well. My focus on how social, political, and economic context figures in the way a technology is built, however, provides an opportunity to consider how the politics might have evolved or the architecture designed if more attention had been paid to the relationship between the genes and ovarian and male breast cancer incidence. Would scientists, physicians, the media, and the public have paid as much attention to the gene discoveries? Would the technologies have been built in the same way? Would the technology have generated as much demand?

Defining the Users of Technology

As we shall see throughout this book, differences in the architectures of genetic testing for breast cancer in the United States and Britain, in terms of the types of clinical care and laboratory analysis that were used and the way testing systems were organized, had important implications for the identities of the individuals interested in using the technology, for the way risk and disease was defined, and for the organization of genetic medicine. The idea that technologies shape those who use them has been explored by many sociologists of technology. Steve Woolgar, for example, has noted that technologies "configure" users while Madeline Akrich suggests a more dynamic interaction between users and technologies, arguing that technological objects articulate a "script" for the identities—behavior, interests, skills, and motives—of users.[54] "Like a film script," Akrich notes, "technical objects define a framework of action together with the actors and the space in which they are supposed to act."[55] These scripts articulate identities for the users of technologies, but do not determine them. As with the technologies Akrich describes, the architecture of each BRCA-testing system in the United States and Britain had quite specific implications for the roles of the actors participating in the system—including their rights and responsibilities, and their power and authority in relation to one another.

We shall see how the identities defined by BRCA testing, like the technologies themselves, were connected to nationally specific ideas about the practitioners and users of health care. The toolkits that social actors used to envision and build their testing systems also articulated user identities based on national histories of and approaches to health care. In the United States, for example, one provider built a technology that characterized the person interested in testing as an empowered consumer, tapping into an increasingly common image of the user of twenty-first-century health care in the United States. As we shall see throughout the book, defining the potential user as a "consumer" or a "patient' is simultaneously the achievement of a specific technological infrastructure, and is connected to broader national ideas about the users of health care.

Indeed, there is really no neutral term to appropriately describe the users of health care. It is for these reasons that I refer to the person who engages with the testing system and is interested in having her blood analyzed for mutations in the BRCA genes as the *client*. I do this because calling her a patient, a research subject, a citizen, a blood sample, a consumer, or even an individual presupposes a certain set of roles, rights, responsibilities, and relationships with other participants within a national context that I argue are actually produced through the processes of assembling a system of genetic testing for breast cancer. Even the word *user* is inappropriate, because both health-care professionals and those interested in having their blood analyzed for gene mutations "use" the test in some fashion. Moreover, the client is not always a user, because she does not necessarily choose to undergo counseling and have her blood analyzed.[56] She is also not simply a blood sample, because the way she is identified and treated is often connected to broader understandings of what it means to be a patient, research subject, or a citizen in a particular country. Even her status as an individual, as opposed to a member of a family, is influenced by national approaches to public health and by concerns about maintaining confidentiality within a private health-care system. Finally, as we shall see throughout the book, it is sometimes difficult to define the client by simply using one of the traditional descriptors listed above; she often fits many of them, simultaneously becoming a patient and a research subject, or a blood sample and a consumer, or a patient and a citizen.

Architectures Defining Risk and Disease

Earlier in this chapter I described the concerns of many commentators that genetic testing will have important implications for the definition of risk and disease, creating, for example, a class of at-risk individuals who will become the subject of both ongoing medical interventions and social discrimination. The idea that technologies shape our understandings of risk and disease has been discussed beyond the domain of medical genetics. In his history of blood diseases in the early twentieth century, Keith Wailoo argues that the diagnostic technologies that were available and their professional and institutional context played a pivotal role in how these diseases were defined. He argues that "technologies have been one of many factors *constituting, creating,* and *complicating* diseases in our time."[57] As we peer inside BRCA testing, we shall learn that it is not simply the presence or absence of a technology that modulates the creation and definition of risk and disease, but also its specific technological architecture. A technology can produce multiple understandings of risk and disease, depending on its architecture and national histories and philosophies of health care. The construction of individuals with genetic mutations as

"presymptomatically ill,"[58] with genes becoming symptoms of new patholo-
gies, for example, is tied to specific testing systems and is not necessarily a
natural result of contemporary excitement about genetic medicine.

Structure of the Book

I begin by investigating the elements of the national toolkits available for
the development of BRCA-testing services in the United States and Britain.
In chapter 1, I compare the two national systems of health-care provision
and show how their approaches to health care led to the creation of different infrastructures for genetic testing in the late twentieth century. In
America's privatized health-care system, genetics services were largely
unregulated, and were available as specialized services at academic medical
centers and through private clinics and laboratories. Individuals often had
a choice of how they wanted to receive these services. In Britain, where
the government guaranteed health care to all citizens, the National Health
Service eagerly developed genetics services as a potentially cost-effective
form of preventive care. It built regional genetics clinics that administered
both counseling and testing services and collaborated with general practitioners in determining an individual's overall care. Individuals were guaranteed access to the genetics services that were provided by the NHS, so
long as their need was certified by a health-care professional. These services were more seamlessly integrated into an individual's health care than
in the United States, where individuals worried that informing primary-
care physicians and health insurers of the results of genetic tests would
eventually lead to genetic discrimination and to loss of health
insurance.

In chapter 2, I begin to explore the development of BRCA testing, comparing the various architectures of the new technology that emerged in the
United States and Britain in the wake of the BRCA gene discoveries. In the
United States, government advisory committees, scientific and medical
organizations, and patient advocacy groups, concerned that the technology would be offered by a private laboratory as a commercial service,
argued that BRCA testing had unclear benefits and risks and should be
offered either in the context of research or as a specialized clinical service.
By contrast, their British counterparts, who had more trust in their
government-run genetics services and were accustomed to lobbying for

increased access, defined BRCA testing as a life-saving technology that should be widely available.

In building their testing services, some US and British providers heeded the recommendations of these groups; others had their own ideas about how the new technology would best be built. All the organizations that began to develop BRCA testing, however, clearly incorporated elements that were specific to their national context. The four providers that initially emerged in the United States replicated the diversity of genetics services in the country, providing BRCA testing as an experimental laboratory technique, clinical research, reproductive services, and a commercial product. In Britain, two types of testing services emerged, reflecting tensions about how best to achieve the goals of the NHS. Initial BRCA-testing services provided by NHS genetics clinics were directed by regional health authorities, but many health-care professionals and public health officials advocated for a national strategy that they felt would better fulfill NHS ideals of equal access to health care across the country.

Not only were both the laboratory and clinical aspects of the BRCA-testing systems built differently according to national context and each developer's objectives, but each provider also clearly defined its appropriate use. One provider in the United States, for example, offered BRCA testing as a commercial product that provided genomic information using state-of-the-art techniques. Here, the user of the testing service was a consumer who had the right, even the obligation, to demand access to DNA analysis as long as she could pay for it. The national testing system in Britain, by contrast, used an integrated clinical and laboratory service to identify, counsel, and manage individuals at low, moderate, and high risk for breast cancer. Individuals who used this system were simultaneously citizens and patients, with the right of equal access to health care but forced to heed the triage recommendations of her physician.

Despite the initially diverse environments of BRCA-testing services, one provider eventually dominated in each country. In chapter 3, I investigate the narrowing of this field of testing services and argue that each provider became successful by adopting strategies that were particularly credible and legitimate in its national context. In the United States, Myriad Genetics used its financial strength and intellectual property rights to force other providers out of the testing market. In Britain, proponents of a national BRCA-testing strategy pressured regional genetics clinics to adopt their system by appealing to their NHS commitment to all British citizens. Not only did the client—and her blood—have a different experience, but the mechanism by which one testing system became dominant in each country also differed quite considerably.

In chapter 4, I argue that the way BRCA testing was integrated into medical care in the two countries highlights different definitions of a good health outcome. Myriad Genetics focused attention to the results of DNA analysis, arguing that its sophisticated laboratory techniques could generate information about the presence or absence of genomic information that was pivotal for a woman who wanted to take charge in decisions about her health care. It also suggested that its test was the only diagnostic tool available to identify a new disease: inherited susceptibility to hereditary breast and/or ovarian cancer. The company further emphasized the importance of its test when it capitalized on the US Food and Drug Administration's approval of tamoxifen as a chemopreventive measure. Tamoxifen quickly became the "cure" for the "disease" of inherited cancer susceptibility that BRCA testing could diagnose. Proponents of a centralized NHS strategy, unlike Myriad, focused on using the test as an additional tool to help identify all British women who were at risk for breast and ovarian cancer, not just those with BRCA mutations. They were reluctant to define new risk and disease categories, instead incorporating the insights of the new genetic analysis into existing understandings of familial risk for the disease. This sense of responsibility to all British citizens was also evident in the reluctance of NHS officials and British health-care professionals to prescribe tamoxifen as a chemopreventive agent; They suggested that, in order for a drug to be offered to a large population of women who, except for being at high risk for breast cancer, were healthy, much more research was necessary.

Once it had become the sole provider of BRCA testing in the United States, Myriad Genetics tried to expand its service to Britain. Its attempt was not successful. In chapter 5, I argue that the elements of the national toolkits that were incorporated into US and British BRCA-testing systems made them particularly difficult to transport across national borders. Myriad Genetics' attempt at international technology transfer drew vehement opposition from British health-care professionals, scientists, patient advocates, and government officials, who challenged the company on the very grounds that had made the company successful in the United States.

They argued that in a country that defined the roles of the health-care professional and the client, the relative importance of laboratory and clinic, and the assignment and use of intellectual property rights quite differently, Myriad's BRCA-testing service simply wouldn't work. Resistance to Myriad was so strong that the company eventually relented, giving up its plan to shut down the NHS's BRCA-testing service.

In the conclusion, I draw lessons from this comparative study for the future of genetic medicine, suggesting that this analysis raises issues that should be a part of our discussions about how to regulate not only genetic testing but also other genetic technologies, including pre-implantation genetic diagnosis and pharmacogenetics. I also discuss how comparative case studies of technological architectures can be useful for the purposes of policymaking, and technology assessment in particular. Finally, I reflect on the implications of nationally specific approaches to science and technology for efforts toward globalization.

In the epilogue, I explore how differences in the US and British approaches to BRCA testing have grown into larger debates in Europe about the appropriate development of genetic medicine, the logic and consequences of gene patenting, and the implications of a commercial scientific environment for the public's health.

1 Toolkits for Genetic Testing

Genetic medicine developed quite differently in the United States and Britain over the latter half of the twentieth century, a result of different health-care systems and approaches to commercialization in biomedical research and technology. In the United States, health care was and continues to be a largely private affair, with doctors and hospitals providing services that are paid for mostly through insurance companies. The government attempts to universalize coverage by offering services to the poor and elderly through its Medicare and Medicaid programs. Still, studies show that more than 40 million people have inadequate or no health insurance.[1] In Britain, the state assumes the direct role of providing medical care for the entire population, through the National Health Service. While private insurance is available, genetic testing for medical purposes has historically only been offered through the NHS.

In this chapter, we shall see how these differences in the provision of health care, coupled with different approaches to university-industry relationships and specifically, the emergence of the biotechnology industry, have led to very different histories of genetic medicine in the two countries. Although medical genetics has very similar origins in the United States and Britain, with early institutionalization of specialized genetics clinics in hospitals and controversial eugenics movements that sought to use hereditary information to control the genetic makeup and health of future populations, their histories began to diverge in the 1960s as the field began to grow beyond these origins. Different philosophies of health care and approaches to the commercialization of research findings led genetic medicine in different directions in the two countries and eventually, created very different toolkits to be employed in the development of BRCA testing.

United States

The United States has had a health-care system based on private insurance since the 1930s. While countries in Western Europe adopted compulsory health insurance and later national health services programs, America's entrepreneurial history and decentralized government structures made it both unlikely and difficult for a national health-care system to emerge. Furthermore, when such ideas were first discussed, American physicians fought the creation of any such system, fearing that it might place considerable power in the hands of government and thus reduce their autonomy.[2] By the middle of the twentieth century, then, when genetics was becoming a recognized medical specialty, private health-care provision in the United States was well entrenched. Clinics and hospitals that began to build genetics services did so of their own volition, without interference from the government.

The first genetics clinics emerged in the first half of the twentieth century in hospitals connected to universities (known as academic medical centers) and hired geneticists for both research and clinical purposes. The field gained further legitimacy in 1948, when the American Society for Human Genetics, the first professional body devoted to medical genetics, was formed. During this time, the typical genetics consultation was usually convened when an individual or her health-care professional suspected a hereditary component to a particular condition or disease present in a family member. While some of these clients would consult with pediatricians or obstetricians, most of them seeking information about whether to marry or reproduce would eventually meet with specialists in medical genetics, discuss their family history, and learn about the hereditary dimensions of their condition or disease. Although geneticists were medically trained, they operated quite differently from most physicians of the day; not only did they not have quick cures for genetic diseases, but they also opted to take a non-directive approach in their consultations. Trying to distinguish themselves from an earlier eugenics movement in which governments, health-care professionals, and scientists sought to take control over reproductive decisionmaking, geneticists did not prescribe a specific course of action, but rather tried to simply provide information and discuss options, about the meaning of an individual's family history of diseases and the consequences of particular marital and reproductive decisions.[3]

These advisory services changed significantly in the 1960s and the 1970s, as laboratory services to detect chromosomal and DNA anomalies were developed and abortion was legalized.[4] More and more diseases could be diagnosed (and some even treated) biochemically, leading to the addition of laboratory to clinical services. Biochemists, for example, developed karyotyping techniques, which allowed geneticists to see the number, form, and size of an individual's chromosomes, which they could use to identify conditions that resulted from extra or missing chromosomes, including Down's Syndrome and various sexual anomalies. Scientists also improved laboratory protocols to test for genetically caused inborn errors of metabolism, including phenylketoneuria and sickle-cell anemia, whose effects could be mitigated with early diagnosis and treatment. Amniocentesis, which detected genetic abnormalities in the fetus through extraction and analysis of amniotic fluid, became available. Clinicians could now use laboratory services to offer a more refined diagnosis of genetic disease as well as more "treatment" options—pregnancy termination, dietary changes, as well as the future possibility of cures. Geneticists remained the primary advisors, gathering family-history information and counseling individuals about particular genetic conditions, but laboratories at academic medical centers now played an important role as well, confirming or rejecting diagnoses made using family-history information.

As genetic medicine began to expand and gain force, the US government took notice. Many states developed newborn screening programs for diseases such as phenylketoneuria and sickle-cell anemia, and in the early 1980s, the President's Commission for the Study of Ethical Problems in Medicine and Biomedical and Behavioral Research, a federal advisory group, was asked to investigate this new area of medicine, articulate the scientific, medical, ethical, and legal challenges it might raise, and make recommendations. Its 1983 report Screening and Counseling for Genetic Conditions suggested that genetic medicine be developed with explicit attention to five ethical principles: confidentiality, autonomy, increasing knowledge, respect for well-being, and equity.[5] As it defined these principles, the President's Commission acknowledged that this new area of medicine would raise regulatory questions; in a country with a private health insurance system, it was likely to be rapidly commercialized, for example, and would thus require the development of professional and quality standards. Its suggested regulatory strategies, however, differed for laboratory and clinical care. Laboratories, the President's Commission recommended, should be subject to federal licensure and proficiency testing, while the physicians providing clinical care should be monitored by their peers in professional societies. Physicians would escape serious government oversight and would offer care as they saw fit.

The commission did acknowledge, however, that additional regulation of both clinics and laboratories might be necessary if commercialization and expansion of services took place haphazardly. "When a screening test is promoted by a laboratory or offered independently by physicians rather than as part of a coordinated program, overall responsibility for coordinating and assessing its availability and quality may be overlooked. . . . Some states have created bodies to oversee the execution and evaluation of genetic screening programs and to avert harm that can result when responsibility for coordinating programs is not clearly assigned. These organizations benefit from both public and professional input in policymaking."[6] Genetic testing, the President's Commission argued, was best when laboratory and clinical services were offered in the coordinated manner that was emerging in hospitals across the country, and disruption of such coordination might prompt increased government involvement. Although the President's Commission's recommendations did not result in any new policies at the time, their assessments regarding the possible commercialization of testing would prove to be quite prescient.

While the President's Commission felt that genetic medicine would require the development of a novel and comprehensive regulatory framework, it seemed quite cautious in its policy suggestions. Why? Three possible reasons come to mind. First, perhaps the President's Commission was reluctant to over-regulate an emerging area of science and industry. American policymakers have historically been quite cautious about regulating the development of science and technology, particularly in the area of biotechnology, citing both the importance of academic freedom and the contribution of innovation to economic growth and national competitiveness.[7] As we shall see below, this consideration certainly influenced later regulatory decisions about genetic testing. Second, it is possible that the President's Commission focused on regulating laboratories because it would be much easier to monitor seemingly objective indicators such as error rates and positive and negative test results than the quality of a health-care professional's counseling techniques. In fact, the United States really had no mechanism to regulate the quality of clinical care; while the British NHS had direct control over the activities of the health-care professionals that it employed, the United States had no analogue. Finally, the President's Commission's reluctance to suggest federal oversight over geneticists might be explained by the longstanding autonomy that American physicians have held throughout most of the twentieth century. As the sociologists Paul Starr and Andrew Abbott have observed, physicians have worked very hard to establish their expertise and avoid monitoring or control by the government, creating authority that had previously allowed them to stop the creation of a national health-care system.[8] In fact, in the United States, while professional organizations such as the American Medical Association write guidelines for how to engage in clinical care, physicians are not required to comply with the recommendations of these groups.

As the President's Commission predicted, genetics did grow rapidly in the 1980s, as researchers began to find genes and build DNA-based genetic tests for diseases (including sickle-cell anemia, cystic fibrosis, and Huntington's Disease). Most of these tests, like the karyotyping and biochemical techniques that had been developed earlier, were initially developed in and provided by research laboratories at academic medical centers. As tests became available for more and more common diseases, however, larger diagnostic laboratories across the country began to recognize that the demand for genetic testing was likely to increase and considerable revenues might be available for those who provided genetic testing on a large scale. Large laboratories at academic medical centers that already offered a variety of diagnostic tests, such as the Mayo Clinic in Minnesota and Baylor College of Medicine in Houston, began to develop infrastructures to offer DNA-analysis services to clinics beyond their institutional walls. Private companies and stand-alone laboratories began to develop such services as well. The Genetics and IVF Institute (GIVF), a private reproductive services clinic founded in 1984, provided not only in-vitro fertilization and egg and sperm donation, but also began to offer prenatal testing for both biochemical disorders (e.g., congenital 21-hydroxylase deficiency) and genetic conditions (e.g., cystic fibrosis). Genzyme Genetics, a start-up biotechnology company based in Boston began offering genetic testing for cystic fibrosis and other inherited disorders in 1986. Such services could have been useful to academic medical centers that did not have scientists conducting research on those particular genes or that did not want to build up infrastructures within their diagnostic laboratories to conduct particular tests (perhaps because they felt that the number of test requests they received would not justify such development). Economies of scale worked to the advantage of these larger clinical laboratories: while most academic medical centers might be reluctant to build up services to test for each rare genetic condition, such services might be more lucrative for large clinical laboratories that were able to test samples from all over the country. Meanwhile, as many of these private concerns became involved in genetics research, they began to apply for patents immediately after finding a gene and as they built testing services. They were then able to use the patents to control commercialization, availability, and the shape of the technology, and thereby maximize their revenues.[9]

Indeed, as genetics research and testing were commercialized by genomics companies, private clinics, and large laboratories both within and without academic medical centers, it became increasingly important for these services to become revenue generators. Providers tried to encourage uptake of their technologies in a number of ways. Some advertised their services through direct mailings to genetics clinics and smaller laboratories at academic medical centers, at professional meetings, and in the *American Journal of Human Genetics*. (See figures 1.1 and 1.2.) Many also emphasized the sophistication of their laboratory techniques or the availability of a package of various genetic tests, offering benefits that smaller research or diagnostic laboratories could not provide.

As these services grew, and as laboratory services began to commercialize, the shape of genetic testing began to change. The laboratory and the clinic, for example, were becoming discrete and independent service providers. No longer were genetic counseling and testing services coordinated under one roof—a patient could be seen by a geneticist in Seattle and her blood tested by a laboratory in Boston, with only the transfer of DNA and paperwork in between. The emergence of stand-alone laboratories and organizations seeking to commercialize their gene discoveries through exclusive development of laboratory services, coupled with the proliferation of genetic tests available, also threatened to challenge the unique power of the geneticist and the value of her specialized services. Previously, clinical geneticists were able to maintain de facto control over access and clinical care related to genetic testing largely because of

Figure 1.1
Advertisement for the DNA diagnostic laboratory at GeneScreen (*American Journal of Human Genetics*, 1989).

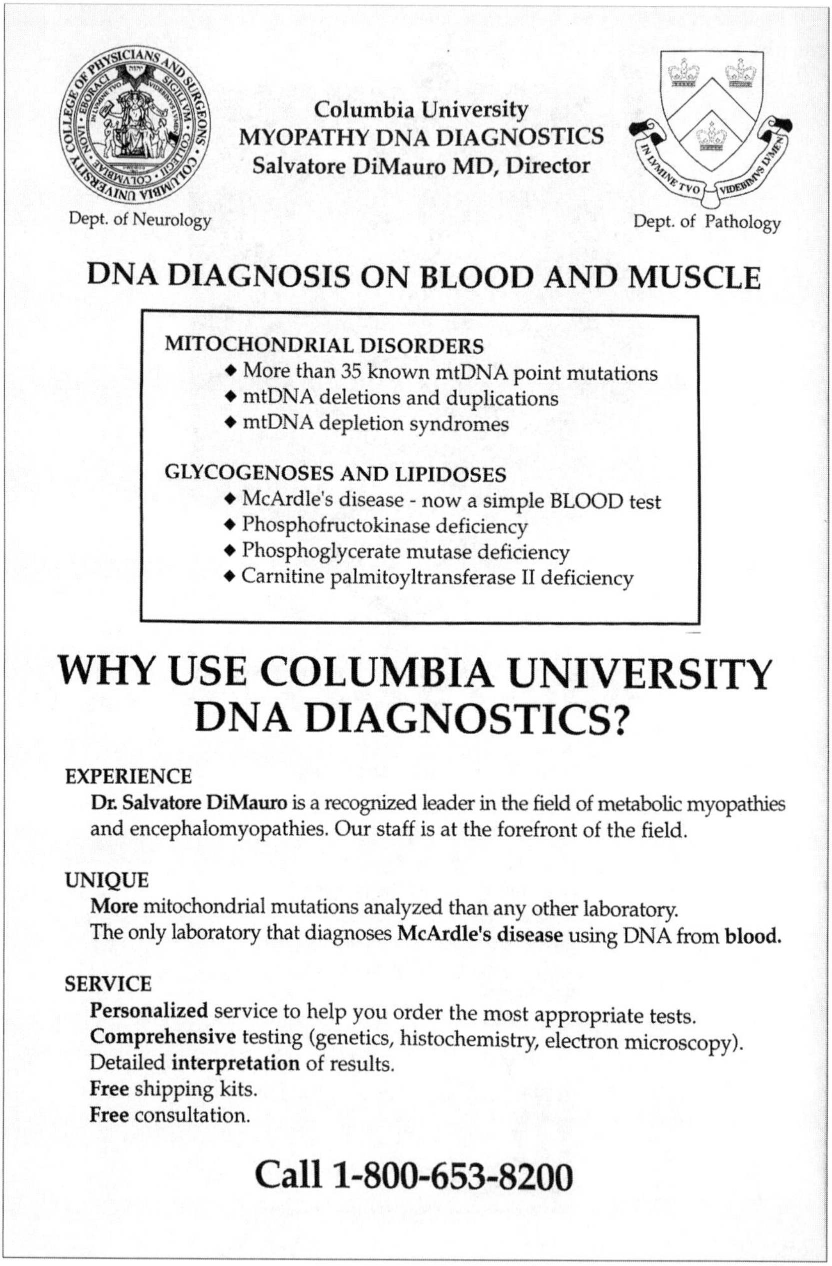

Figure 1.2
Advertisement for the Myopathy DNA Diagnostics Laboratory at Columbia University (*American Journal of Human Genetics*, 1996).

the small scale of services and connection to the laboratory. However, as more people wanted to use these services, and as laboratories began to offer services independently, there were no restrictions preventing other clinicians, including primary-care practitioners, from offering genetic counseling or helping clients get access to DNA-analysis services. Although the American Society of Human Genetics and the National Society of Genetic Counselors tried to dissuade counseling by non-geneticists by emphasizing their own unique expertise in position papers, articles, and participation in national advisory committees,[10] the demand that was being generated by public discussions about the power of genetics could not be contained by the few genetics clinics scattered across the country. As laboratory and clinical services split apart and counseling moved out of the proprietary domain of the geneticist, genetic testing itself began to shift in focus. No longer was it a system coordinated between the specialized clinic and the laboratory, it was now a DNA test that could be offered through any physician. These changes, however, did not escape government attention.

Novel Regulatory Challenges

When genetics services first began to expand in the late 1980s, they were largely fitted into existing regulatory frameworks. Despite the extensive public discussions throughout the country about the impact of genetic medicine, very few new policies were devised to deal with this set of technologies in terms of its laboratory or clinical dimensions or its social or legal implications. The President's Commission report had had little regulatory impact, so genetics services were automatically treated like all other laboratory tests and subject to federal control through the Clinical Laboratory Improvement Act of 1967 (CLIA 67), which covered any laboratory that provided more than 100 tests per year in interstate commerce for profit. Administered initially by the Centers for Disease Control (CDC) and the Health Care Financing Administration (HCFA) and eventually by the Center for Medicare and Medicaid Services, CLIA had mandated monitoring and inspection systems for laboratories and laboratory personnel. The act, however, did not cover most genetic testing laboratories, specifically academic research laboratories, that offered a relatively small volume of tests. It also lacked an enforcement mechanism, and HCFA officials would learn that a laboratory had not complied with the CLIA 67 regulations only if that lab was the subject of a complaint.

A few states and non-profit organizations stepped into this regulatory void. The state of New York, for example, not only developed mandatory standards that covered all personnel and laboratories offering genetic testing, but also required New York clinicians to use only those laboratories that it had approved. In order to gain this approval, laboratories had to demonstrate that they successfully used accepted test methodologies and undergo proficiency testing. Inspectors of these laboratories had to undergo a special training program as well. Meanwhile, non-profit organizations such as the Council of Regional Networks of Genetic Services and the College of American Pathologists developed voluntary quality-control programs with which laboratories could certify themselves, and then perhaps use the certification to market the quality of their services and thereby increase test volume. The National Tay-Sachs and Allied Diseases Association, which used such a strategy, widely publicized the names and contact information of the laboratories that had been approved by the International Tay-Sachs Disease Quality Control Reference Standards and Data Collection Center, distributing the list of accredited laboratories to Jewish organizations around the world. In fact, their system led to the closure of some laboratories and improvement of services in others. As one might expect, these monitoring and enforcement mechanisms, which were often voluntary, were quite uneven across tests and states. Furthermore, while these rules increased oversight over laboratories, they did not investigate the distance that was emerging between genetic laboratories and clinics or the clinical dimensions of testing, demonstrating a focus on the laboratory aspects of genetic medicine that would eventually play an important role in the development of BRCA testing.

In 1988 the Clinical Laboratories Improvement Act was amended to expand federal oversight to include almost all laboratories that conducted an "examination of materials derived from the human body for the purpose of providing information for the diagnosis, prevention, or treatment of any disease or impairment of, or the assessment of the health of, human beings."[11] The new regulations (known as CLIA88) also developed standards for scrutiny and approval based on whether a test was deemed to be of moderate or high complexity. While almost all genetic tests were now technically under CLIA's purview, very little specific attention was specifically paid to this category of tests. Genetic tests were still, despite the considerable public interest in and attention to them, considered in regulatory terms as similar to other laboratory tests. Not only were genetic tests not classified as a subspecialty that required focused attention and targeted guidelines, but the complexity of very few genetic tests had been specifically defined. Perhaps most importantly, many research laboratories could still escape the purview of CLIA altogether—personnel at these small laboratories might not know that they were subject to CLIA regulations, or they simply might not be known to the central laboratory at their institution or to CLIA regulators. In addition, the focus remained on the laboratory—moderate-complexity and high-complexity tests, for example, were determined according to the type of gene to be analyzed, not the type of counseling required.

By the early 1990s, as genetics seemed poised to expand far beyond its origins as a small specialty with the impending discovery of genes linked to cancer and Alzheimer's disease, national advisory committees were convened to again investigate the issue broadly, in terms of the social, medical, and policy challenges that the new area of medicine raised. First, the NIH's National Center for Human Genome Research funded the Institute of Medicine, part of the National Academy of Sciences complex, to study the "current status and future implications of such testing." As Stephen Hilgartner has observed, NAS committees are often called in to investigate important policy issues because they are seen as groups of "qualified experts" who can "deliberate privately in an unbiased manner about issues of vital national importance."[12] The IOM committee, which was made up of experts in law, molecular and medical genetics, religious studies, bioethics, health policy, and medicine, issued its report, titled "Assessing Genetic Risks," in 1994.[13] The report, emphasizing that these new technologies were at a critical stage of development and should be considered investigational until their benefits and risks had been def .ed, suggested that attention be paid to both the laboratory and clinical dimensions of genetic tests. It suggested that CLIA be extended to explicitly cover genetic tests, and that control, monitoring, and proficiency testing measures be strictly enforced. The report also recommended that all genetic tests be regulated by the US Food and Drug Administration, and that they be placed in "Class III," a category of medical devices that requires full pre-market approval. Class III medical devices must be submitted for pre-market approval because insufficient information exists to assure safety and effectiveness, and are usually those that support or sustain human life, are of substantial importance in preventing impairment of human health, or which present a potential, unreasonable risk of illness or injury.[14] As it suggested that the FDA's infrastructure for regulating medical devices could be extended to cover genetic tests, the IOM committee characterized genetic testing as more than simply laboratory analysis. It suggested that the new technology needed an investigational period and was connected to both laboratory and clinical services. It wasn't enough, the committee argued, to determine whether a laboratory did its work properly; some clinical dimensions of the test—the relationship between mutation and disease incidence, for example—were part of the testing system and thus subject to regulation. In the case of genetic testing, review of both the laboratory and clinical dimensions would require FDA to assess analytic sensitivity (probability that a test will be positive when a particular gene sequence is present) as well as clinical sensitivity (probability that positive test results are correlated with people with the disease or people who will get the disease, depending on the gene mutation and disease in question). Tests could be made available on an investigational basis within research protocols that had been approved by Institutional Review Boards—institutional committees devoted to ensuring that research projects involving human subjects were safe and ethical—while widespread availability would be dependent on FDA approval. The IOM committee was thus suggesting that genetic testing was a medical device, and that both its laboratory accuracy and its quality for clinical purposes should be assessed before it was made widely available.

In its report, the IOM also articulated its interest in the clinical dimensions of genetic testing by warning that a variety of clinicians—not only geneticists—would soon have to learn how to counsel individuals curious about their genetic risk: "Since genetics education and counseling are likely to be provided increasingly by primary-care practitioners, these practitioners will need training to help them perform these functions appropriately and to know when to refer patients to specialized genetics personnel."[15] While the committee acknowledged the unique role of clinical geneticists, it also suggested that all clinicians would have to increase their expertise in medical genetics. In particular, it recommended that medical education be revised to include information about the technical, ethical, social, and legal issues involved in genetics. Rather than abandon the specialized counseling that had been offered almost exclusively in the domain of the geneticist, the committee suggested that all physicians be trained to fulfill the important counseling elements of the technology.

The IOM argued that it was particularly important to train all physicians to offer genetic counseling when dealing with the genetics of late-onset disorders, as tests for such diseases were useful only to a small fraction of people, a lack of gene mutations (even for someone with a family history of the disease) did not mean that the person would never get the disease, and genetic testing was only likely to provide variable probabilistic information. While there would be considerable uncertainty about the meaning of mutations in susceptibility genes, however, there would likely be considerable public interest in the discovery of genetic "causes" for diseases such as cancer and heart disease, and thus the IOM committee was extremely concerned about how susceptibility gene information might be used, for example, to make medical, insurance, and employment decisions. The committee also worried that because these technologies were likely to generate considerable public excitement, and private providers might be tempted to offer them widely and quickly to take advantage of demand, special care should be taken to ensure proper provision and use. "Strict guidelines for efficacy," the report stated, "will be necessary to prevent premature introduction of this technology."[16] If genetic testing were to be made available without adequate oversight, the committee warned, the information generated by the new technologies might be misunderstood by health-care professionals, employers, and insurers, and might have negative effect on a client's health care.

While most of the IOM committee's recommendations with regard to the role of the FDA and appropriate regulatory strategies were not implemented, Centers for Disease Control, through its CLIA Advisory Committee (CLIAC) that was responsible for advising and making recommendations on the technical and scientific aspects of CLIA, responded to the IOM report by convening a working group to decide whether a genetics subcommittee should be created and whether special attention should be paid to this category of tests. In addition, in true Washington fashion, the National Institutes of Health–Department of Energy Working Group on Ethical, Legal, and Social Implications of Human Genome Research followed the IOM committee's recommendation to convene another advisory committee to study genetic testing in further detail. In 1995, the Working Group convened a Task Force on Genetic Testing (TFGT) to "review genetic testing in the United States and, when necessary, make recommendations to ensure the development of safe and effective genetic tests."[17]

In contrast to the IOM membership roster, which represented the various areas of academic expertise relevant to deliberations on genetic testing, the TFGT brought together members of the various groups that might be considered stakeholders in the development of genetic testing. The committee included representatives from the major scientific and medical professional organizations, insurance industry, biotechnology industry, NIH's Ethical, Legal, and Social Implications of the Human Genome Project Research Program, and patient advocacy groups. In contrast to the IOM, which gained its legitimacy from the academic pedigrees of its committee members and the closed nature of its deliberative processes, the TFGT tried to establish the importance of its recommendations by emphasizing how well its committee represented the interested parties.

The TFGT met throughout 1996 and part of 1997, hearing testimony from experts in various fields and soliciting comments from members of the public. The concurrent development of genetic testing for breast cancer was also pivotal to the TFGT's deliberations and recommendations. As members discussed the appropriate provision and use of genetic testing, they witnessed the excitement over the discoveries of the BRCA genes and eventually the introduction of testing services. In keeping with the way the new technology was being viewed by scientists, physicians, and bioethicists, it also became a test case for the TFGT, and it was characterized as such in the text of the final report, published in the fall of 1997.

The TFGT agreed with many of the IOM committee's observations and recommendations. Its report argued that genetic tests, and particularly predictive genetic tests for common diseases, posed novel ethical, social, and medical challenges. Not only would these genetic tests offer risk information that was likely to be useful to only a small number of people (and not even necessarily useful to all those at high risk for a given disease), it suggested, but also very few medical interventions were available for those who were defined as genetically "at risk." The TFGT also echoed the IOM committee by noting that public excitement about the availability of genetic tests for common diseases would likely encourage rapid commercialization of the technologies and involvement of non-geneticists (such as primary-care practitioners and surgeons) in the provision of counseling services and explanation of ever more complicated risk/benefit ratios.

Like its IOM predecessor, the TFGT recommended that all genetic tests initially be considered investigational and clinically assessed through protocols approved by Institutional Review Boards. This was particularly important, the committee argued, because genetic tests were being increasingly commercialized and were subject to premature availability. If organizations wanted to introduce their tests to the wider market as early as possible, the TFGT suggested that the FDA could grant conditional premarket approval, which would allow developers to market their test while making clear that "safety and effectiveness were still under investigation."[18] This investigational period would then be devoted to investigating clinical validity and clinical utility, not just analytic validity, which was already covered by CLIA. Clinical validity included clinical sensitivity (probability that the test would detect a gene sequence if it was present), clinical specificity (probability that the test would be negative when a particular gene sequence was absent), and predictive value (probability that a person with a positive test result has or will get the disease for which a particular gene sequence is used as predictor), while clinical utility meant that the genetic test could be used to improve health outcomes. By suggesting that the FDA needed to assess clinical utility, the TFGT suggested that the government needed to have even more oversight responsibilities than the IOM committee had recommended. Overall, the TFGT demanded a comprehensive review of both the clinical and the laboratory dimensions of genetic tests before they were made widely and commercially available. They defined a genetic test as more complicated than a Class III medical device. It was a novel technology that required a new regulatory scheme.

The TFGT also lauded the additional oversight of laboratories that would occur if CLIAC created a genetics subcommittee, and argued that national proficiency testing and on-site inspection measures be strengthened. It again took a slightly stronger approach than the IOM, however, arguing that laboratory directors must have formal training in human and medical genetics and that *all* predictive genetic tests (those that provided clients with risk probabilities) should be designated as "high complexity" and thus subject to stricter scrutiny. Overall, while it agreed with many of the IOM's recommendations for the regulation of genetic testing, the TFGT provided a much more detailed and comprehensive framework than the IOM committee and suggested that FDA extend its oversight in an area that had previously only been regulated by CLIA.

In particular, the TFGT went beyond the IOM committee's recommendations by suggesting that the FDA should judge not only the laboratory accuracy of the test, but also its utility for disease prevention. It suggested that the architecture of a genetic test was not simply a collection of laboratory practices and machines, but also included clinical care such as how information was transmitted to clients and what types of medical interventions had been devised to deal with at-risk status. While the FDA "does not generally assess safety and effectiveness of a laboratory test in terms of its ability to improve outcomes of those undergoing testing [clinical utility],"[19] leaving physicians to interpret results and make recommendations about clinical care, the task force argued that the physicians offering testing would increasingly be non-geneticists who might not have the specialized knowledge to appropriately offer testing and interpret results. Thus, the TFGT argued, the FDA should step in to ensure not only that the tests were safe and effective, but also that clients were receiving correct information to help them make better health-care and lifestyle decisions. These issues needed to be addressed in the case of genetic testing not only because of the psychological implications for individuals who knew their gene mutation status but could do nothing about it, but also because many genetic tests were becoming available for which no medical interventions had been devised. Here, the TFGT went far beyond the recommendations of previous advisory committees and suggested that both the laboratory and clinical dimensions of genetic testing should be covered by FDA regulation. While it did not suggest that counseling could only be offered by specialized geneticists, it did emphasize that physicians with particular expertise needed to offer care in conjunction with genetic testing—thus promoting the integrated approach that had been supported, more subtly, by the IOM and the early President's Commission. This recommendation, however, also threatened to diminish the authority of physicians by suggesting that regulators were in a better position to determine who had the expertise to provide care and whether a test was medically beneficial and should be made available to clients. In order to help the FDA deal with these specific challenges posed by genetic tests, the report suggested, it should convene a panel of consultants to guide its regulation of these new technologies.

The TFGT also suggested specific interventions to improve physicians' knowledge of genetics and improve their preparedness for genetic counseling. The problem of offering genetic counseling to a public that was being informed about genetic tests through advertisements and the media, the committee noted, could not be solved simply by learning the appropriate textbook definitions of medical and molecular genetics, discovering the rapidly changing state of genetic science and technology, and understanding appropriate and effective counseling strategies: it also required a different style of medical care. Physicians who were accustomed to providing directive advice, recommending one treatment over another, may not be well equipped to provide the non-directive advice that had been developed over the past decades of medical genetics. "Another drawback," the report stated, "is the tendency of non-geneticist providers to be directive in situations in which reproductive options to avoid the conception or birth of an infant with a serious disorder are considered. . . . because of past efforts to deny people the opportunity to reproduce because they possessed presumably heritable traits, and the need to respect personal autonomy in reproductive matters, efforts to steer people towards a particular reproductive decision are undesirable."[20] The TFGT recommended better genetics training in medical schools and even the inclusion of genetics questions on licensure and specialty board certification exams. It also suggested that hospitals and managed care organizations develop mechanisms to ensure that providers of genetic services, both specially trained geneticists and non-geneticists, were competent to provide counseling. It was also particularly important to increase public education in genetics in an era of commercialized testing, the report noted, and it supported the development of new models of genetic education and counseling.

While the TFGT deliberated and developed its recommendations, both the US Senate and House of Representatives held hearings during 1996 to assess the social and regulatory challenges posed by genetic testing. Government regulators, however, seemed unlikely to take on the additional responsibilities that would soon be detailed by the Task Force. At the House hearing, for example, Mary Prendergast, deputy commissioner of the FDA, stated:

At present, we estimate that there are dozens of companies and laboratories that are now offering hundreds of genetic tests to the public, and we predict that this number will grow rapidly. Any new proposals to regulate, whether by the FDA or by anyone else, this rapidly growing, important technology have to take into account the issues and concerns of increased costs and the potential stifling of innovation and investment into consideration.[21]

Not only did it worry about its regulatory capacity within its own budget, but the FDA was concerned about how best to protect the public's welfare. It implicitly asked the fundamental question about how to regulate new areas of innovation—should it develop a strict regulatory framework to control use of the new technology or maintain vigorous innovation with less regulation?

Perhaps not surprisingly (in view of the FDA's stated position), the TFGT's report had a lukewarm reception. The only recommendation to be immediately adopted was the creation of yet another advisory committee, this time chartered by the Department of Health and Human Services. In June 1998, the Secretary's Advisory Committee on Genetic Testing, which was similar in composition to the TFGT, began to "assess, in consultation with the public, the adequacy of current oversight of genetic tests and, if warranted, to recommend options for additional oversight."[22]

As genetic testing for breast and ovarian cancer was built in the 1990s and the 2000s, there were no major changes to the oversight of the laboratory or clinical dimensions of genetic testing. While the CLIA 88 regulations still covered genetic testing laboratories, plans to develop a genetics subspecialty within CLIAC stalled, and genetic testing was still available separately as laboratory and clinical services and from a variety of sources. In the late 1990s and the early 2000s, molecular geneticists worked with CLIAC to develop regulations covering analyte-specific reagants, the active ingredients of diagnostic tests, which were increasingly available for sale to laboratories setting up genetic-testing services. Of course, these proposed regulations still focused entirely on the laboratory aspects of the test. Meanwhile, physicians (sometimes through professional organizations) continued to oversee themselves and make their own determinations about the clinical utility of genetic tests.

The growth of genetic medicine in the United States during the second half of the twentieth century provided developers with quite a varied toolkit as they began to build BRCA testing, with some elements contradicting others. In keeping with traditions of entrepreneurship and the philosophies of health care envisioned by a private medical system, genetic testing was becoming a commercial enterprise, with laboratory and clinical services increasingly offered in separate venues. Regulators were reluctant to involve themselves in the development of genetic testing in any way. They worried that they might stifle a growing industry, and perhaps they assumed that a genetic test, because it was not invasive like a drug, could not be dangerous. In addition, both the regulatory attention to the laboratory dimensions of genetic testing as well as early advertisements for DNA-analysis services by both academic and industrial providers focused attention to the activities of the laboratory as the essence of testing. It would be entirely reasonable, then, if BRCA test developers built it as a proprietary laboratory technology. There was, however, as there is in the rest of American health care, some ambivalence about the role of the private sector, as advisory committees worried about the separation of counseling and laboratory services, a focus on DNA results themselves rather than their interpretation, the expertise of non-specialists offering genetics services, and the consequences of providers' profit motives. Although these regulators and advisors had little effect on policy, and there was little government control over the provision of the new technology, we will see in the next chapter that their arguments and attempts to shape the debate over genetic testing—emphasizing, for example, the need to integrate counseling and laboratory services—influenced the considerable diversity among the BRCA-testing services that were initially built in the United States.

Britain

The Benefits of Genetics

In contrast to the market environment that had emerged in the United States, genetic testing in Britain was built entirely by the National Health Service, and the way these specialized services were built reflected the history and the politics of the national health-care system within which they emerged. State involvement in British medical care began early in the twentieth century, in parallel with other Western European countries who adopted national social welfare programs that included compulsory health insurance, primarily in an effort to provide acute and emergency care to low-income workers to promote industrial productivity.[23] Britain initially developed a national health insurance program through its 1911 National Insurance Act, which guaranteed free care from general practitioners to workers who earned less than a certain income. Over the next decades, the government gradually became more responsible for controlling not only the health care of low-income workers but also the administration of most hospitals. By the 1940s, the government had decided to create a national system to organize and oversee all of these services, and in 1948 the National Health Service was born. The Health Minister at the time, Aneurin Bevan, described the new system as a "comprehensive health service to secure improvement in the physical and mental health of the people . . . and the prevention, diagnosis and treatment of illness."[24] Funded through taxation, the system originally guaranteed access to the health-care system through a network of twelve regions that governed day-to-day physician and hospital care and a national office which set broad policy and provided funding. Although the creation of the NHS initially engendered resistance from British physicians similar to that in the United States, Bevan is said to have used financial incentives to convince them to join the new system, a strategy that was particularly successful because their salaries had been relatively low.[25]

The NHS's goal of providing the British population with equal access to health care has been interpreted in a variety of ways since its creation. In the 1970s, the NHS aligned local and regional health-care authorities in order to ensure that every individual in every locality was linked to the full spectrum of health-care services and to make the lines of responsibility for health-care provision clear from the national administration to every city and town. As part of its general drive toward privatization in the 1980s, the Thatcher government tried to improve health-care services by injecting some market mechanisms in the NHS structure. It created a decentralized internal market, where regional health authorities competed with one another to get money from the central NHS administration to pay for all health-care services, including public health programs and specialized services such as genetic testing. The Thatcher administration argued that the internal market would allow physicians to focus on their patients by shifting their administrative and bureaucratic duties to administrators at the regional health authority.[26] The system separated the provider and purchaser of health care, and created a managed market with physicians competing with one another for regional funding and regions competing with one another for money from the central government. Although the Thatcher government argued that the initiative was intended to improve patient care, however, many critics felt that the new system focused simply on the availability of services rather than the quality of those services.[27] Others noted that the competitive atmosphere for health-care services led to the provision of very different services among regions, and thus contradicted the NHS's goals of equal access to health care across the country.

When Tony Blair became Prime Minister, in 1997, most plans for an internal NHS market seemed to disappear and the national administration of the NHS was strengthened. In fact, the Blair government revised a Patient's Charter that had been initially written in 1991 to emphasize national health-care standards and the *equal* rights and responsibilities of patients across *all* NHS regions.[28] As we shall see throughout this chapter and the rest of the book, this history of NHS politics had serious implications for the structure and provision of genetics services in Britain.

Throughout the second half of the twentieth century, as NHS priorities and organizational hierarchies changed with the goals of different national administrations, genetic medicine was growing. In Britain, as in the United States, genetic services re-emerged in the 1950s and began to expand in the 1960s and the 1970s with the development of amniocentesis and newborn and fetal testing technologies. Obstetricians and pediatricians offered prenatal and newborn screening services while NHS-funded teaching hospitals began to build genetics clinics and develop testing infrastructures in their laboratories for both clinical and research purposes. These structures for provision were thus somewhat analogous to the clinics housed in academic medical centers in the United States, except that they were entirely under the purview of the NHS.[29] As they evolved, however, British genetics services began to look quite different from their counterparts in the United States, reflecting the government's public health goals and the priorities of various national political administrations. Depending on the politics of the time, genetics services were administered by NHS officials at local, regional, and national levels, which led to variation among the technologies and services that were available and to whom.

British geneticists initially justified expansion of their services within the NHS during the 1960s and the 1970s, Peter Coventry and John Pickstone have observed, by arguing that their services would be particularly helpful as the NHS tried to cut costs and improve health care through streamlined services and increased attention to preventive care.[30] As teaching hospitals were brought under the control of NHS regions, the genetics clinics that were housed within them had the potential to become a vital and integrated part of health care. If funded properly, they could serve as a central location for the early detection, prevention, and treatment of disease. Geneticists also noted that they, not pediatricians or obstetricians, were best equipped to provide genetics services because they could understand the meanings of family histories of disease and counsel individuals and their families about genetic conditions and the treatments options available to them. Not unlike their counterparts in the United States, as British geneticists sought to institutionalize genetic medicine within the NHS, these health-care professionals tried to stabilize a unique role for their profession as well.

NHS officials considered the arguments of their geneticists, specifically those that referred to the preventive and cost-saving benefits of their services, quite seriously. In a series of "Prevention and Health" pamphlets issued after it had streamlined services and created clear lines of national administration in the early 1970s, the NHS made genetics services an important part of its national prevention strategies. The first pamphlet in this series, "Prevention and Health, Everybody's Business," published in 1976, accepted that prenatal diagnostic testing could have significant preventive benefits: ". . . rapid advances have been made in techniques to diagnose abnormal conditions of the fetus with the object . . . where these are serious enough, of considering abortion. In one sense this amounts to prevention since otherwise the outcome would be the birth of a grossly affected individual."[31] It also recognized possible cost-saving benefits: "The possibility of detecting disease at an early stage by the use of a cheap easily-administered screening procedure is an attractive one. Where such procedures can economise on the use of scarce skilled staff then their advantages would seem even more evident."[32] Not surprisingly in this country with a government-run health system, genetic medicine was immediately considered in terms of its population-wide benefits, both in terms of disease prevention and cost savings for the NHS. Though the President's Commission had also considered the population benefits of national (or even state-based) screening programs, cost saving was not considered as significant an issue in a private insurance-based system, and there had been much more of a focus on the risks and benefits of the new technologies for individual clients.

Although it is logical to assume that the NHS would be more inclined to accept technologies that could have preventive and cost saving benefits, officials still recognized the potential problems involved in offering prenatal diagnostics when limited treatments except for selective abortion were available. One pamphlet in the "Prevention and Health" series noted: "For many of the abnormalities being considered, e.g., Down's Syndrome, there is no cure; and the choice lies between abortion or the birth of an affected individual. How should the community weigh the interests of an individual yet unborn?"[33] While these sorts of questions had also been raised in the United States, the British government's overall attitude toward these services had a much greater impact on their use because it operated as both a regulator and provider of health services. As discussed earlier in this chapter, in the United States, regulators had not interfered with the availability of these technologies despite the passionate politics of abortion, and individual providers—innovators, hospitals, and physicians—could choose whether and how to offer these services. In Britain's government-run NHS, officials at either the national or regional level decided whether to fund services at all and whether to offer auxiliary services such as counseling or support in dealing with the birth of affected children, and thus could often control how services were shaped. Indeed, individual physicians and hospitals had limited autonomy as administrators could choose to stop funding services that they felt were poorly developed or structured.

Another major concern of NHS officials as they tried to determine how to shape these new technologies was how to achieve the NHS goal to provide all citizens with equal access to them. This goal presented a fundamental dilemma. Testing everyone for a genetic disease would be cost-prohibitive, but only testing high-risk individuals might be too limiting and violate the principle of equal access. An NHS report stated:

> . . . for many diseases there are groups of the population which are at higher risk than others. If value for money were the sole consideration one might wish to restrict preventive and screening programmes to these high risk groups which will give the greatest yield for a given expenditure of money. But this aim may conflict with the principle that everyone should have equal access to medical care if they can possibly benefit from it. And yet if everyone is to have equal access to medical care and the funds available are limited then the preventive programmes which can be carried on will be less and the amount of avoidable illness or premature death will be greater. There is no easy solution to this dilemma and the balance which will be struck is bound to be one which will not satisfy everybody.[34]

Risk assessment and triage were already important elements of the NHS's toolkit. As Ashmore, Mulkay, and Pinch have described in their analysis of British health-care economists, development of rational standards such as triage, used to guide medicine and distribute access to services in an equitable manner, is a recurrent tool used in the National Health Service.[35] While triage methods seemed to conflict with goals to offer equal access to medical care, the NHS suggested that equal access did not mean unlimited access, and that care should be based on need rather than demand. In time, genetic information could also potentially facilitate triage by offering additional biochemical markers that could refine decisions about how treatment was offered and to whom. Triage on the basis of physiological need, as we shall see throughout this book, was used to distribute BRCA-testing services as well, although both the way it should be conducted and how equal access was defined was the subject of much discussion and controversy.

By 1977, with the publication of "Prevention and Health, Reducing the Risk," the NHS seemed to have completely accepted geneticists' arguments about the benefits of their services, articulating its enthusiasm for these new technologies and considering appropriate implementation strategies. More than 20 genetics services had been built in hospitals across the country, and the pamphlet emphasized the importance of the specialized counseling services that the clinics could provide. It also argued that laboratory services should be improved and extended to match genetic counseling, so that an integrated service could be provided to citizens: "Laboratory facilities for prenatal diagnosis have developed unevenly over the country and there is a clear case for rationalization of this service on a regional basis to meet present and future needs more effectively. Such facilities will need to be extended to match the growing demand for genetic counseling."[36] The laboratory and clinical dimensions of genetics services were slowly being woven together and becoming an increasingly central and stable part of the NHS's approach to genetic medicine; in fact, for these British services, the clinic, rather than the laboratory, seemed to be the central part of the testing system.

As more genetic tests became available throughout the 1980s, the laboratory and clinical aspects of genetics services in Britain continued to expand together within the NHS. More NHS regions, which were directly in charge of these services, built genetics clinics and developed laboratory infrastructures for genetic testing. Chromosomal and DNA testing were provided entirely within the NHS, either by research laboratories at teaching hospitals or by regional diagnostic laboratories. When a new gene was discovered, testing would initially be provided by a research laboratory until the regional laboratory had developed the infrastructure to offer it or had contracted with another regional diagnostic laboratory to provide services. Some commercial laboratories began to offer DNA-analysis services; however, they worked solely with regional genetics clinics rather than doing business with general practitioners or other specialist physicians, thus maintaining the integration of specialized genetics counseling and DNA analysis. While these services had begun to fragment in the United States during this period, with private laboratories and clinics starting to offer genetics services separately and widely, in Britain there was a clear emphasis on keeping both laboratory and genetic counseling services within the NHS, emphasizing the specialized expertise of genetics clinics, and maintaining an integrated relationship between the laboratory and the clinic. Of course, in Britain, where the vast majority of genetics services were being purchased by the NHS, there was also little incentive for private clinics or laboratories to try to increase demand or competition in this market where the government had so much control over pricing.

Genetics in the Policy Domain

As in the United States, the growth of genetics services in Britain caught the attention of government officials far beyond the health sector, and in the mid 1990s, the House of Commons Science and Technology Committee considered a range of issues related to human genetics and its social, political, ethical, and legal implications. The Parliamentary inquiry lasted more than a year, during which it heard evidence from geneticists, physicians, patient representatives, representatives from the insurance industry, lawyers, clerics, social scientists, bioethicists, and officials from the Department of Health and Patent Office. The commission issued its report, titled "Human Genetics: The Science and Its Consequences," in early 1996, recommending the creation of advisory committees covering the field of human genetics, and genetic testing in particular, in order to deal with the novel social, ethical, legal, and psychological issues raised by the new science and the new technologies. While these advisory committees would not have statutory authority, they would be able to consider developments in the field and make recommendations to government departments and Parliament if they felt that legislative interventions were necessary.

Prime Minister John Major's administration responded to the Select Committee by initially creating the Advisory Committee on Genetic Testing (ACGT) and eventually the Human Genetics Advisory Commission (HGAC), both in 1996. The memberships of the HGAC and the ACGT seemed to combine the approaches of the IOM and the TFGT, by including individuals on the basis of their research expertise and status (moral philosophers, lawyers, medical and molecular geneticists, experts in medical sociology, bioethics, and rehabilitation studies) as well as their stakeholder interests (representatives from the biotechnology and pharmaceutical industry, genetic support groups, and the government). Both the HGAC and the ACGT had journalists on their rosters, from the British Broadcasting Corporation and the *Times* of London, respectively. The director of science communications for the Science Museum in London also sat on the ACGT. Why were these members chosen to sit on the committees? Two possible reasons come to mind. First, the committees had an interest in communicating directly with the citizenry, and these members could bridge the gap between science and the public. Second, whereas in the United States patients were viewed as critical consumers who needed to participate in health policy decisionmaking, in Britain's government-run health-care system the press (the customary watchdog and critic in policy matters) was the appropriate participant. As we will see, however, this began to change through the BRCA-testing episode as patient advocates became more active and powerful in Britain. Although one might easily assume that all advisory committees convened by governments might have the same purpose, we see that they can have multiple missions and that these missions are reflected in the way their memberships are put together and also, perhaps, in the type of advice that they provide.

The HGAC, which reported to the Departments of Health and Trade and Industry, was created to examine the non-health-care areas of human genetics, including insurance and employment, intellectual property, and privacy and discrimination issues. Soon after its creation, it issued a report on the potential for discrimination in life insurance on the basis of genetic information and became embroiled in discussions about whether and how results of genetic tests should be used in insurance underwriting.[37] It also issued reports about cloning and the use of genetic-test results in employment decisions.[38]

The ACGT, which was administered through the Department of Health, was created to oversee the provision and use of genetic testing and address any regulatory challenges that might arise.[39] Responding to an attempt by a private company to offer cystic fibrosis testing outside the context of NHS genetics clinics or even a primary-care physician, the ACGT's first task was to develop guidelines for the provision of genetic testing directly to the public. It recommended that all such services first undergo an approval process, and that these tests be "limited to determination of carrier status for inherited recessive disorders *in which an abnormal result carries no significant direct health implications for the customer*" [emphasis added].[40] As it argued that the availability of such tests should be limited to non-health-related testing, the Committee demonstrated its commitment to the genetic testing system developed by the NHS through its regional genetics clinics. Although the ACGT did not have any statutory authority, organizations familiar with the US context who were seeking to commercialize genetic medicine, and particularly health-related genetic testing in Britain, would be quite disappointed with its position. As it eschewed commercial provision of genetics services, the report also expressed national contentment with the NHS approach to genetic medicine. This sentiment and allegiance to NHS-sponsored genetics services, we shall see, would play an important role in the way British BRCA testing was built.

The ACGT's second report, which focused on services of genetic testing for late-onset disorders such as breast cancer, reiterated the committee's commitment to the NHS's integrated counseling and laboratory services and in many respects resembled the IOM and TFGT reports issued in the United States. It described the scientific, medical, social, and ethical issues raised by genetic testing for these diseases and emphasized that as requests for genetic testing increased, it was particularly important to ensure the quality of both clinical care and laboratories. It suggested, for example, that physicians who were not genetics specialists would have to become more proficient in understanding genetic etiologies: "Because of the increasing use of genetic testing by all clinical disciplines it is important that the skills learnt by many clinical geneticists, not only in relation to collecting and analyzing genetic data but also such general aspects as empathy, information giving, acknowledgement of family issues and confidentiality, are taught within the medical school curriculum."[41] While the ACGT's report encouraged increased genetics proficiency among all medical professionals, however, it still saw the integrated laboratory and counseling services offered by the regional genetics clinics (which had grown to 27 in number by the mid 1990s) as the central locations for genetic medicine in the United Kingdom. Perhaps responding to the proliferation of private genetics laboratories and increasing distance between laboratory and clinical services in the United States, the report recommended: "Genetic testing should be undertaken only by laboratories closely linked with other genetic services. . . . Although the technology may be the same [as other diagnostic tests], genetic testing for inherited disorders, in particular pre-symptomatic testing, requires different approaches, and should not be undertaken by general laboratories unless they form part of or are closely affiliated to a genetic testing service. . . ."[42] As it took this position, it continued to validate both the institutionalization of genetic medicine and the professional identities of laboratory and medical geneticists. Even as genes were linked to more common diseases and genetic testing became more widespread, regional genetics clinics and their affiliated laboratories had an important and distinct role to play.

By the mid 1990s, genetic medicine in Britain had emerged quite differently than it had in the United States. It was originally incorporated into the national health system because of its preventive and cost-saving benefits, and was being built using elements that had were common to the NHS, including risk assessment and triage, and with regional centers as tertiary sites of care that were connected to national administrators through regional health authorities. Services remained entirely within the NHS, and counseling and laboratory services were offered in an integrated fashion by the NHS's regional genetics clinics.

Conclusion

The environments of genetic medicine that emerged in the United States and in Britain, and thus the toolkits that were available to developers of BRCA testing in the two countries, looked quite different by the time of the BRCA gene discoveries in the mid 1990s. In the United States, genetic testing was available from a variety of sources—university research laboratories and clinics, gene discovery companies, private clinics and diagnostic laboratories—and in a number of forms. BRCA test providers could choose to offer their services as part of research protocols or commercial technologies, as a test that could be administered only by a specialist or through any physician. While government advisory committees and other critics continued to advocate increased regulation of the field and monitoring of commercial services, consumers bought genetic technologies in a competitive marketplace where the involvement of health-care professionals and responsibilities of test providers varied quite considerably. Not only was genetic testing not subject to any additional regulation, it was not even monitored under existing FDA regulations on drugs, medical devices, or direct-to-consumer advertising.

In Britain, genetic-testing services were offered within the NHS, through its regional genetics clinics, with an integrated approach to testing and counseling. In addition, providers often used triage and risk assessment to distribute services as they did with many other medical interventions within the NHS. There was limited room for variation, because of the coordinating and funding role played by the central NHS administration. In these two countries, despite similar interests in genetic science and technology, different approaches to health care and commercialization of research seemed to be leading genetic medicine in divergent directions.

2 Comparative Architectures of Genetic Testing

Debates in the United States and Britain about how best to build genetic testing for breast cancer began as soon as discoveries of the BRCA genes were announced. Should it be offered in an integrated manner, or should laboratory and clinical services be provided separately? Should it be prescribed by primary-care physicians, or was it part of specialist care? What should be done with the test results? Through articles in scientific journals, press releases, position statements, and interviews with the media, a variety of groups—including patient advocates, scientists, health-care professionals, and prospective testing providers in both countries—began to answer these questions. As they tried to influence this next step in the development of genetic medicine, however, they did much more than simply express their opinions. They proposed specific "architectures" of genetic testing for breast and ovarian cancer—including, as I described in the introduction, system components and ways of fitting them together—that defined appropriate roles for those engaged with the technologies.

This chapter investigates and compares the multiple BRCA-testing architectures proposed in the United States and Britain in the mid 1990s. In both countries, interested individuals and groups had conflicting ideas about the appropriate shape of the new genetic testing technology. However, each of the architectures that were envisioned incorporated elements of the national toolkits that had already begun to be assembled to shape genetic medicine. Indeed, we shall see that although there was variation within each country, BRCA-testing architectures were clearly products of their national context and differed markedly between the United States and Britain. In the United States, a private health-care system, a largely unregulated approach to genetic medicine, a history of commercializing biotechnology and genomics, and a tradition of patient activism played important roles in how the tests were built. In Britain, by contrast, the historical provision of genetics services by the National Health Service as well as the overall politics of the health-care system—including debates about how best to provide equal access to health care across the country—guided the way test developers built their new technologies. Starting with the American technologies and then moving to the British, I will discuss the various testing systems that were envisioned in the two countries. As I introduce each system in turn, I will describe how elements of national toolkits were incorporated into the technology and then follow a hypothetical client as she journeys through the system. As I describe each journey, I identify the consequences of each system's architecture for its participants.

United States

Within the diverse and minimally regulated genetic testing environment of the United States, multiple interpretations of the nature, purpose, and appropriate shape of BRCA testing arose as groups considered how best to build the new technology. Should it be added to the existing menu of services provided only by specialized genetics clinics? Or was it an ordinary medical test, like those measuring cholesterol levels and blood pressure, which could be administered by any physician? Three types of groups— breast cancer activists, professional organizations representing health-care professionals and scientists, and prospective test providers—took the initiative in offering answers to these questions and offered rather different visions of how BRCA testing should be developed.

Breast Cancer Activists
The Washington-based National Breast Cancer Coalition and the San Francisco-based Breast Cancer Action (by the mid 1990s the most influential advocacy groups involved in breast cancer politics at the national level) cautioned against the widespread availability of the new technology. They worried that the risk information generated by BRCA testing would provide ambiguous results, because of the risk, rather than certainty, of future disease incidence and the paucity of medical management options available. Therefore, both suggested that testing be offered in a highly regulated manner and only in conjunction with extensive clinical care. In fact, the

NBCC went so far as to say that testing should only be provided through research protocols.

Representatives of these patient advocacy groups began to express such opinions almost as soon as the BRCA gene discoveries were announced. In a front-page *New York Times* article announcing the BRCA gene discovery, Nancy Evans, president of BCA, noted: "It's a very mixed blessing to have this knowledge . . . it's the first step in a long journey, and the journey is probably across a minefield."[1] Five days later, Fran Visco, president of the NBCC, worried: "Women will have to be very careful. . . . You're talking about giving them a test telling them they have an 85 percent chance of getting a disease that we don't know how to prevent, and for which there is no known cure."[2] Both women, and the groups they represented, urged the public to see the gene discoveries as simply another step in an ongoing research process rather than a solution to the breast cancer problem.

The opinions of these groups were particularly important. Following in the footsteps of the women's health movement of the 1970s and the AIDS movement of the 1980s, breast cancer activists had risen to prominence in the early 1990s by demanding greater influence over US government policies related to research and treatment for the disease. By the mid 1990s, breast cancer activists increasingly assumed expert and advisory roles in the media and on government advisory committees, weighing in on each new advance in treatment, health-care controversy, and proposed change in research funding. One member of the NBCC observed: "I think for the Coalition, I just think that we have a much more reasoned, analytic way of looking at problems. And I think we have, I know we have the respect of many people on the Hill, when they have a breast cancer issue, they call the Coalition to see what we have to say."[3] The development of genetic testing for breast and ovarian cancer was no exception, and breast cancer activists quickly supplemented their initial reactions in the media with policy statements that described their visions of the new technology.

A Client's Journey through the Breast Cancer Activists' System
A client using the systems proposed by the NBCC and BCA would first visit a health-care professional who could counsel her about the benefits and risks of BRCA testing. In its policy statement published in a bimonthly newsletter that reached more than 2,000 subscribers, BCA stated: "No one should be tested without access to education and counseling concerning

all benefits and risks of genetic susceptibility testing. . . ."[4] Neither organization specified where this counseling should take place or how it should
be conducted. They did not specify, for example, that such counseling be
offered by a specially trained geneticist and thus did not take an explicit
position in the debate that had been raging for years over whether non-
specialists in genetics could provide counseling. However, one might easily
assume that both imagined a counseling session that included information
about the limitations and implications of the test for clients concerned
about cancer risk and their families.[5]

The NBCC suggested further that clients have access to testing only
through a research protocol. Its position paper recommended: "Because
much more needs to be researched about the sensitivity, specificity, and
reliability of the genetic tests and because not enough is known about the
effectiveness of genetic education and counseling, genetic testing should
only be available within peer-reviewed research protocols."[6] These studies
usually investigated the psychological impact of testing or the utility of
the technology for disease prevention and/or management, and were also
likely to be run by specialists in genetics. These specialists would likely
provide specialized counseling and facilitate their research subjects' access
to DNA analysis. The NBCC recognized, however, that provision of the
technology through research protocols would also likely limit access to
clients with a personal or family history of the disease. The NBCC dealt
with this restriction, which conflicted with one of the organization's stated
goals to "improve access" to health care, in two ways.[7] First, it suggested
that research would eventually benefit all those concerned about their
BRCA risk by gathering additional information about the validity and
utility of the test. Second, it recommended that research protocols be available widely, so that individuals across the country could get tested and participate in the investigative process. "It is imperative," the NBCC stated,
"that such research studies should be made available to those for whom
such testing is appropriate and ultimately that such studies should be
widely available, easy to access in both urban and rural areas."[8]

Regardless of whether counseling took place in a research or clinical
setting, both groups recommended that the client and her health-care professional decide together whether to pursue testing and send blood to the
laboratory for analysis. Neither the NBCC nor BCA made specific recommendations about the laboratory procedures that should be used to analyze

the BRCA genes, but both argued that commercial laboratories should not be used for these services, expressing similar caution to many of the advisory committees who had earlier reviewed the role of the private sector in genetic medicine. The NBCC noted that testing on a commercial basis was inappropriate when the "reproducibility, sensitivity, specificity, and predictive value of tests" were unknown, and BCA pointed to the profit motives of companies who might jeopardize the well-being of clients by making the technology available prematurely in order to capture the market. Both suggested that people only had to have their BRCA genes tested once in a lifetime to know whether they contained mutations, and that if clients were tested before adequate research had been conducted on the relationships between specific mutations and disease incidence they might make inappropriate conclusions based on the results.[9]

After the genes were analyzed, the client returned to the clinic on an ongoing basis for post-test counseling and long-term follow-up.[10] Both groups were careful, however, to note that both counseling and laboratory services should remain confidential unless the client decided otherwise. BCA urged that "the array and number of unresolved issues related to genetic testing for susceptibility to breast cancer make compelling the need for written informed consent prior to such testing or to the release of the results of such testing to third parties."[11] This confidentiality of genetic-test results was already standard practice at genetics clinics in the United States, who had responded to concerns that genomic information in the medical record could fall into the hands of insurers or employers and cause discrimination.

Defining the Roles of System Participants

As the discussion above suggests, the architecture of the breast cancer activists' proposed BRCA-testing systems defined specific roles, rights, responsibilities, and authority for system participants. They described women as citizens and patients who had two primary rights: to choose among good medical options and be protected from bad medical choices. For both BCA and the NBCC, commercially available testing was considered a bad medical option. The NBCC limited access further by suggesting that research protocols were the only good medical option available. Meanwhile, BCA argued that testing could either be offered in the research or clinical setting but must be accompanied by education and counseling. It also emphasized that BRCA testing was an inadequate answer to the breast cancer problem:

It is equally clear that the BRCA1 test for genetic susceptibility is not the early detection tool we need . . . a positive result from the BRCA1 test does not mean that the person tested will develop breast cancer. (Nor does a negative test mean she is not at risk.) And, even if a positive test meant a woman would certainly develop the disease, there is currently no known effective method of preventing breast cancer.[12]

Who would be responsible for recognizing good medical options and protecting individuals from bad ones? Activists argued that this task should fall to health-care professionals, the US Food and Drug Administration, and themselves. Health-care professionals, they felt, had a duty to ensure that clients were properly educated and counseled about the benefits and risks of testing. In addition, the FDA should assert its authority and not allow commercial testing without adequate research. Meanwhile, activists would continue to define themselves as appropriate authorities with the expertise to distinguish between good and bad medical options. In the newsletter article described in the preceding paragraph, for example, BCA displayed its ability to distinguish between good and bad science as it methodically detailed reasons why testing was dangerous. The NBCC also identified itself as an expert in the definition of good medicine: *"Together we* [emphasis added] can make certain *we* get the data we need. Too many medical recommendations in breast cancer—on how to treat women, what tests to give them—are made without a basis in good science. *We* must not add genetic testing and its followup to this category."[13] Not only was commercial BRCA testing not in the category of good science, activists argued, but clients were not necessarily in the position of deciding what types of health care were best for them. Instead, they recommended, clients should be advised and protected by the state, physicians, and knowledgeable activists about the appropriateness of particular health-care options. Like the AIDS activists Steve Epstein describes in his research exploring how the AIDS community gained power in biomedical policymaking, these activists were using their scientific and medical expertise to prove their authority to speak for the needs of women.[14]

As they advocated limited choice to BRCA testing and presented themselves as authorities in distinguishing between good and bad science, breast cancer activists also distanced themselves and their expertise from the individuals they represented. They claimed a combination of scientific training and expertise in the patient experience that authorized them to distinguish between health care based on good science and a technology that was potentially harmful. Meanwhile, however, they argued that their constituencies did not have a mastery of scientific knowledge and needed to be protected by them as well as physicians, test providers, and the government. Epstein mentions this phenomenon briefly in his analysis of AIDS activists, noting that one stated: "I *never* represented 'people with AIDS.' I represented *activists*. And those are different people, you know. They are a subset of people with AIDS."[15] As breast cancer activists asserted their ability to distinguish between good and bad science and their authority to determine how the new technology should be provided and used, they also distinguished their unique knowledge and became self-appointed experts.

Empowering the Client

The efforts of breast cancer advocacy groups to construct a testing system that provided women with a limited right of access may sound surprising considering that breast cancer activists had a history of encouraging women to take charge of their health care and that they operated among other patient advocates, particularly in the United States, who lobbied for greater access to innovative medical care. However, the breast cancer activists' attempts to protect women from the BRCA-testing technology was by no means unprecedented. In fact, this episode provides an example of how patient advocates often craft their identities among complicated and multiple definitions of empowerment, protection, and choice.

Women's health groups and disease-focused social movements have long negotiated between objectives of empowerment and protection as they articulate their identities. The first edition of *Our Bodies, Ourselves*, published in 1973, served as the modern, empowered woman's "bible" that launched a generation of women's health activism. It popularized the phrase "Knowledge is Power," and it emphasized the importance of an individual's control over her body through knowledge, particularly in the face of what the authors perceived to be a paternalistic medical establishment: "Finding out about our bodies and our bodies' needs, starting to take control over that area of our lives, has released for us an energy that has overflowed into our work, our friendships, our relationships with men and women, and for some of us, our marriages and parenthood."[16] Simultaneously, however, the book advocated additional government regulation to protect women from dangerous drugs and medical devices. These activists defined empowerment as including greater awareness about one's body and better access to health care as well as protection from dangerous medical interventions. Such a definition allowed women's health activists to demand increased access to research funding for women's health issues while blaming the government, and specifically the FDA, for premature approval of diethylstilbestrol (DES) and the birth control pill, both of which had caused serious side effects during the 1960s and the 1970s.[17]

Activists tried to reconcile empowerment objectives with efforts to protect women by qualifying the definition of empowerment to include access to "good" science and medical care, as opposed to "bad" knowledge or technology that could be dangerous and therefore disempowering. As discussed above, many involved in the BRCA-testing episode adopted the same strategy, arguing that women would be empowered through controlled access to BRCA testing. Indeed, although breast cancer activists focused on "empowering people to deal with the issues raised by a breast cancer diagnosis,"[18] the NBCC advocated access only to BRCA testing through research protocols, and BCA stated "we should be a long way from offering a test to anyone who wants it."[19] As these activists distinguished between good and bad science and medical care, they performed what the sociologist Thomas Gieryn has called "boundary work," defining good science in such a way that it reinforced their own ideology and expertise.[20]

There were, however, feminists and women's health advocates who disagreed with BCA and NBCC's definitions of empowerment vis-à-vis BRCA testing. Some activists, for example, argued that women would only truly be empowered through unfettered access to genetic testing. The NIH's Advisory Committee on Research on Women's Health, which was made up of doctors, scientists, lawyers, social scientists, and public health officials primarily concerned with women's health issues, reviewed the availability of genetic testing for breast and ovarian cancer in 1996. Although it initially resolved to restrict testing to the research context and bar unlimited availability, some committee members criticized what they considered to be a paternalistic approach to medical care. Marjorie Schultz, a law professor from the University of California at Berkeley and member of the committee argued: "Can you imagine yourself saying to a woman who comes to a center to do testing, 'No you can't unless you're a research subject?' "[21] The committee eventually recommended, as BCA had, that testing be conducted in the context of counseling rather than recommending that it be restricted to research protocols. Empowerment had a multiplicity of meanings, even among the advocacy community, and each of these definitions had different implications for the provision of medical care and the configuration of rights for clients who engaged with the health system.

Professional Organizations

A number of professional organizations representing scientists and health-care professionals also joined the discussion about the future of BRCA testing, recognizing that the offices of their constituents would soon be full of clients concerned about their risks of contracting breast and ovarian cancer. Like the breast cancer activist community, professional organizations adopted a cautious approach. They lauded the gene discoveries as major milestones in both the study of breast cancer and in genetic research, but emphasized the uncertainty that surrounded any new testing technology and the need for concurrent clinical care. Even among these organizations, however, there was disagreement about whether the technology should be classified as investigational, and who should be eligible to use it.

Four national professional societies proposed systems of genetic testing for breast and ovarian cancer in carefully crafted position statements published after the BRCA gene discoveries. The American Society of Human Genetics (ASHG), representing 8,000 researchers, academics, clinicians, laboratory practice professionals, genetic counselors, nurses, and others interested in human genetics, was the first to issue a statement immediately after the discovery of the BRCA1 gene was announced. Over the next two years, three more organizations issued statements: the American Medical Association, the nation's largest physician's group; the American Society of Clinical Oncology (ASCO), which represents the approximately 20,000 physicians who treat cancer patients; and the American College of Obstetrics and Gynecology (ACOG), which represents approximately 40,000 physicians providing health care for women. The statements of these groups had two purposes. First, they articulated a clear position to influence debates about how the new technology should be built. Second, they defined clear roles for their constituents in the testing system and thus ensured their continued involvement in the development of genetic medicine.

A Client's Journey through the Professional Organizations' System

These groups suggested that clients could gain access to BRCA testing through a primary-care physician, through a specialist, or through a research protocol. Clients using ASHG's or ACOG's system, just as in the NBCC's system, could only be tested through a research protocol (preferably by a genetics professional or obstetrician/gynecologist). Research subjects, these organizations argued, would have access to special protections and better clinical care.

ASCO, by contrast, suggested that health-care professionals (e.g., oncologists, primary-care physicians) use a client's family history to determine access. It asked health-care professionals to recommend testing only if clients had (1) at least two family members with breast cancer and one with ovarian cancer, (2) at least three family members diagnosed with breast cancer under the age of 50, or (3) had or was one of two sisters with breast and/or ovarian cancer under the age of 50.[22] Although this restriction would neither provide the additional protections accorded to research subjects nor enhance broader understandings of breast cancer genetics by contributing to the investigational process, it would be of clinical value to the client. ASCO argued that the limitation would increase the utility of testing and future risk-management options for clients, because the mutations found in those with family histories of breast and ovarian cancer would be more likely connected to future disease incidence.

Regardless of how clients gained access to testing, all organizations agreed that BRCA testing should be approached cautiously in the context of education and counseling. Counseling would include risk assessment based on family history of the disease, discussion of the medical, social, psychological, and ethical issues involved in testing, and potential risks and benefits for the client and her family. Only ASHG specified the appropriate provider of these services: "Clinical geneticists are uniquely qualified to obtain the most reliable information available, to provide a source of continuing information, and to communicate complex ideas and uncertainty in a way that is helpful to the patient."[23] Others simply suggested that their constituencies, with appropriate training, could provide counseling and participate in the integration of genetics into medicine. "Genetic testing," ASCO prescribed, "should . . . be made available to selected patients as part of the preventive oncologic care of families only in conjunction with appropriate patient education, informed consent, support, and counseling. These issues must be addressed by all health-care professionals, whether they be oncologists, genetic counselors, medical geneticists, or primary-care providers, who plan to offer genetic testing for cancer susceptibility."[24] It was, of course, in the best interest of each of these organizations to suggest that their constituents could provide the clinical care that was needed to accompany the laboratory analysis.

If, after counseling, the client decided to pursue testing, her blood would be drawn and sent to the laboratory. Like the patient advocates, these organizations did not specify appropriate laboratory practices, except to emphasize that researchers should continue to refine available analytic methods. They did, however, make some recommendations as to the types of laboratories that should be used for testing. ASCO suggested that oncologists pay attention to "a laboratory's ability to provide accurate, state-of-the-art genetic predisposition testing to at-risk families" beyond the CLIA regulations that monitored the quality of reagents and procedures used in genetic testing.[25] Except for ASHG, who shared breast cancer activists' discomfort, none of the organizations explicitly rejected provision of testing through commercial laboratories.

Breast cancer activists and professional organizations also agreed on appropriate procedures after blood analysis. They recommended the provision of post-test counseling and long-term follow-up care, and advocated that access to the genetic information generated be tightly controlled.

Defining the Roles of System Participants

There were many similarities between the roles of system participants envisioned by patient advocates and those envisioned by professional organizations. Like the NBCC and BCA, scientific and medical groups emphasized the client's right to be protected from bad medical options, including testing without counseling or testing outside the context of a research protocol. They also defined the health-care professional as a protector who had a duty to educate herself about the genetics of breast and ovarian cancer and the benefits and risks of testing, and to care for the client accordingly. Unlike the activists, however, professional organizations saw the health-care professional as taking a primary role in determining who should have access to testing and how it should be provided. The interaction between health-care professional and client envisioned by these organizations resembled the traditional doctor-patient relationship.

Health-care professionals had the responsibility to determine who would have access to testing, and they had a duty to promote the welfare of the client. Meanwhile, although clients did not have an unlimited right to demand access to testing, they could take advantage of the expertise of a specially trained health-care professional in making decisions about whether to undergo testing and, if they tested positive for a mutation, choose what medical management options to pursue.

There was, however, one important difference between the traditional patient and the identity of the new BRCA-testing client. While the doctor-patient relationship was usually considered an interaction between individuals, many of these organizations identified the client as part of a family in which there may be a history of cancer. In fact, ASCO suggested that access to care be limited according to a client's family history of breast and ovarian cancer. As was discussed in the introduction, many scholars have suggested that this transformation of the client from an individual to a member of a family is one of the major new challenges posed by genetic medicine. Diagnoses of genetic risk and disease now have repercussions far beyond the individual client, as family members must take this information into account as they make their own health-care and lifestyle decisions. We shall see, however, that even this implication of genetic medicine depends on the architecture of the testing system.

In the mid 1990s, as breast cancer activists and professional groups proposed architectures for the new technology, a number of organizations began to develop services to test the two BRCA genes for mutations. During 1995 and 1996, at the peak of discussion about the appropriate provision of BRCA testing, four major providers emerged in the United States. The testing services they built differed considerably from one another, with some operationalizing the architectures proposed by breast cancer activists and professional organizations and others drawing inspiration from existing genetic-testing services and other parts of American health care.

The University of Pennsylvania's Genetic Diagnostic Laboratory

The University of Pennsylvania's Genetic Diagnostic Laboratory (GDL), which set up its testing service in 1995, followed many of the recommendations of patient advocacy groups and professional organizations. It offered BRCA testing in the context of its research and required all clients to undergo counseling at an academic medical center.

As a research laboratory at the University of Pennsylvania Health System, one of the country's leading academic medical centers, GDL operated according to the dual priorities of research and health care common among many of the genetic-testing services located at academic medical centers in the United States.[26] These dual priorities of research and health care guided the architecture of GDL's services, and specifically its BRCA-testing system. Since the early 1990s, the laboratory had been developing a cheaper and faster alternative to DNA sequencing called *conformation sensitive gel electrophoresis* (CSGE). In order to refine this technique, GDL provided testing services to the public for a number of rare diseases; it gathered DNA samples from a number of clients, tested them for mutations using CSGE, and then returned the test results. By 1993, GDL researchers had established that CSGE could be a useful alternative to full-sequence analysis of genes.[27] In 1994, they decided to develop a service to test for the BRCA genes, which would provide them with the opportunity to demonstrate the utility of CSGE even further: they could show that their technique could find mutations in long and complicated genes for late-onset disorders as well as simpler genes that had already been found.[28]

GDL, however, offered its BRCA-testing service differently than it had its other genetic tests. While it allowed its other genetic tests to be ordered through any physician, the laboratory required that its BRCA-testing service be ordered through an academic medical center which would provide extensive counseling services. This gatekeeping mechanism was intended to ensure that clients received counseling from a health-care professional specially qualified in genetics, who would explain genetic risk and the benefits and risks of testing.[29] GDL ensured that, even in the absence of government regulation, its BRCA-testing system was consistent with prevailing views in the medical genetics community as well as the recommendations of scientific and professional organizations and patient advocacy groups.

A Client's Journey through GDL's System

GDL did not advertise its BRCA-testing service directly to the public. Instead, clients learned about GDL's system either through their own initiative in contacting an academic medical center directly or through the referral of a physician. Once clients gained entrance to a genetics clinic at an academic medical center, they would meet with health-care professionals (usually medical geneticists or genetic counselors) and receive genetic counseling. While GDL did not involve itself in the clinical inter-action directly, it could easily assume, because the service was being pro-vided by a trained specialist, that genetic counseling would include both risk assessment using family-history information as well as a comprehen-sive discussion of the risks and benefits of testing. In order to conduct the risk assessment, health-care professionals would record a client's family history of breast and ovarian cancer (including the age of onset and details about each cancer) and make a recommendation about whether testing would be beneficial.[30] Depending on the client's family-history and lifestyle information, health-care professionals at the genetics clinic might also offer some the opportunity to participate in a research study that would cover payment for laboratory analysis.

Presented with the probability of a BRCA gene mutation, the client would then decide whether she wanted her blood to be sent to GDL for laboratory analysis.[31] Although health-care professionals never forced the client to undergo laboratory analysis, some refused to test those whom they believed were extremely unlikely to carry a BRCA gene mutation.[32] Some also suggested strongly that a member of the client's family who had already suffered from breast or ovarian cancer be tested first. Because so little was known about the amount of risk conferred by each of the hun-dreds of BRCA mutations, testing an affected family member first could help health-care professionals better understand how specific mutations were related to disease incidence in particular families.

If the client decided to undergo BRCA testing, she (through the genet-ics clinic) would send a blood sample to GDL. In addition to the sample, GDL required that the health-care professional at the clinic and the client send a payment ($700 for testing the BRCA1 gene and $1,500 for testing both BRCA genes) as well as completed forms that documented medical and family history and written proof of consent to the testing procedure. These forms functioned as what Susan Leigh Star and James Griesemer have called "boundary objects," because they operated between two worlds: they distilled the activities of the clinic—counseling, taking of family history, solicitation of consent—into entries on pieces of paper that would provide the laboratory with the information needed to conduct its analysis, including documentation for legal and ethical purposes and evidence of the individual's consent.[33]

Once GDL received the blood sample and other materials, researchers tested it using CSGE, the experimental DNA analysis technique they were trying to refine. According to GDL researchers, the family-history information included with the blood sample helped them determine where to look for a BRCA gene mutation (certain patterns of family history suggested specific mutations, or mutations in a particular gene).[34] A client's family history played an important role in how the DNA analysis was conducted.

Once the blood sample was analyzed for mutations, GDL returned the test result to the health-care professional and the client at the academic medical center. The test result conveyed whether or not a mutation had been found and, if one had been found, what kind of increased risk the mutation likely conferred. GDL could only provide a range of possible risk, however, as not enough data had been gathered about each mutation to provide finer risk probabilities. It was then up to the staff at the genetics clinic to help the client understand the meaning of the test result. GDL was no longer involved. If the test result showed the client positive for a BRCA mutation, staff at the genetics clinic presumably informed her about the meaning of such a mutation in the context of her family history and instructed her about options for managing BRCA risk. The responsibility of clinical management was then left up the client, who decided how to proceed with her clinical care. In fact, as recommended by patient advocates and professional organizations, unless the client specifically requested the communication, neither GDL nor staff at the genetics clinic conveyed the results of the BRCA test to the client's primary-care physician. This was done to prevent notation of the test results in the medical record, which could eventually fall in the hands of insurers or employers. This reluctance to include genomic information in the medical record, however, could impede continuity of care when primary-care physicians lacked information about potentially significant genetic-test results.

Defining the Roles of Participants

Although GDL adopted an approach that was familiar to many academic medical centers in the United States by providing laboratory services to help finance its research goals, it played a greater role than many other academic laboratories in defining who could participate in its experimental BRCA-testing system. Only specialist health-care professionals trained in genetics and employed at academic medical centers could provide clinical care to clients, and clients were likely to be subjected to stricter eligibility criteria because they could gain access to GDL's system only through a specialized clinic. Eligibility was usually determined on the basis of a client's personal or family history of breast and ovarian cancer. In addition, access tended to be limited to those who lived or worked within a short distance of a center (which was attached to a university and likely in a metropolitan area) and were aware of the specialized services it offered. Indeed, GDL restricted access to testing even more than BCA and most professional organizations had suggested. The client, then, was envisioned simultaneously as a part of a family, a patient with restricted access to specialized clinical care, and a subject of DNA research.

Finally, although GDL allowed only health-care professionals affiliated with a genetics clinic at an academic medical center to provide testing services, it did not attempt to manage the interaction between the health-care professional and the client directly. The health-care professional was the ultimate authority for determining whether to recommend testing, to allow participation in research protocols, or to prescribe counseling before and after testing.

OncorMed

OncorMed, a medium-size start-up biotechnology company based in Gaithersburg, Maryland, also offered BRCA testing in the research context. The architecture of its testing service, however, differed considerably from GDL's. It incorporated the priorities of a start-up biotechnology company as it paid attention to the concerns of those in the American patient advocacy and biomedical communities who felt that BRCA testing should only be offered through clinical research protocols. Rather than providing testing in order to refine an experimental laboratory technique, OncorMed's protocols were designed to ensure strict attention to counseling and to limit access to high-risk individuals.

OncorMed—a subsidiary of the molecular biology products company Oncor, Inc.—was founded in July 1993 as a company focused on using genetic discoveries and technologies to improve medical care in the area of cancer. Unlike some of the companies that simply looked for disease genes and developed diagnostics and therapeutics based on these discoveries, OncorMed not only searched for genes linked to inherited susceptibility for cancer and built genetic testing technologies, but also developed medical management tools to help physicians identify individuals who were at an increased risk for cancer and might benefit from the company's technologies. In its 1995 Annual Report, the company stated: "OncorMed provides a valuable linkage between new breakthroughs in cancer genetics and the research and technologies needed to translate these discoveries into diagnostic and therapeutic interventions. . . . We are the gene discoverer's link to one of the world's largest hereditary cancer databases. We are the innovator's link to clinical cancer genetics for promising new technologies, and the physician's link to some of the most sophisticated patient management tools available."[35] The company's first product, the Hereditary Cancer Risk Assessment Service (HCRAS), launched in 1994, consisted of a software package and a professional training program designed to help health-care professionals assess hereditary cancer risks.[36] Health-care professionals could use the software package to collect family-history information from individuals and display a family pedigree that would determine the pattern of cancer incidence in the family. The company expected that eventually, HCRAS would be integrated with genetic diagnostic services, thereby linking technology and medical practice. This involvement of a start-up biotechnology company in breast cancer genetics research and development, particularly at such early stages of innovation, was a uniquely American phenomenon that had become rather common in the development of genetic medicine, as was discussed in chapter 1. Over a number of decades, the United States had developed a network of laws that encouraged start-up companies to accelerate commercialization of research through both venture capitalist funding and relationships with government-funded academic researchers.[37]

In keeping with this overall strategy to integrate state-of-the-art technology with medical practice, OncorMed sought to develop a BRCA-testing service and conducted research to find the BRCA genes. The company also decided to apply for patents on its gene discoveries, following the examples of many others in the United States during the 1980s and the 1990s that applied for and sought to use patents on genes to create a proprietary basis for commercialization.[38] In one of the most famous cases, the NIH scientist Craig Venter and his colleagues applied for patents on cloned pieces of DNA of unknown function that had been created as part of research linked to the Human Genome Project.[39] Although the patent applications stirred considerable controversy and were eventually denied because they did not fulfill the utility criterion set forth by the US Patent and Trademark Office, Venter's attempt popularized the idea that human genetic information could be owned. In fact, after this episode, Venter left the NIH to create Celera Genomics, a company devoted to mapping and sequencing the human genome privately, which joined other companies in selling subscriptions to databases containing human genomic information.

Although it had not been credited with finding either of the BRCA genes, Oncormed applied for patents covering both of them. It had been actively involved in the gene discovery research, using techniques that allowed it to identify a BRCA1 consensus sequence (which was built by sequencing the BRCA1 genes of a number of individuals, and then at the most highly polymorphic—variable—points, averaging the most likely bases to be found at that location to create a full sequence.) This technique generated a different DNA sequence than Myriad's (which was a sequence of an actual BRCA1 gene), and was thus patentable as a separate entity. Multiple patents on the same gene could be granted if they described different gene sequences, which would be entirely possible in the case of the BRCA genes because of their complexity and the multiple mutations possible.[40] In this complex patent environment, OncorMed tried to strengthen its proprietary position even further by amassing licenses on other patents covering the BRCA1 gene (and, eventually, the BRCA2 gene). It purchased a license on a BRCA1 patent held by the geneticist Mary-Claire King and her colleagues, which covered a number of markers on the BRCA1 gene, and also negotiated a license on the BRCA2 gene patent held by Mike Stratton and the UK Cancer Research Campaign, the organization who had funded Stratton's research.

In negotiating the license agreement with Stratton, OncorMed agreed to stipulations that would limit its monopoly power and shape the way the patent would be employed in clinical practice.[41] The agreement required OncorMed to allow the British NHS to use the BRCA2 gene sequence for testing and future research without paying royalties or license fees, committed OncorMed to sublicense the patent to other companies, and not only specified that individuals be counseled before and after testing but also provided a list of procedures that counselors had to follow.

Stratton's counseling guidelines and desire to couple counseling with laboratory testing fit in well with OncorMed's interest in clinical care.[42]

This dedication to influencing clinical services, not simply providing stand-alone testing as offered by many companies for other diseases, was also reflected in the company's choice of Patricia Murphy to build and direct its BRCA-testing service. Murphy was a medical geneticist who was board-certified in both clinical cytogenetics and molecular genetics. She had also served as a member of two federal advisory committees, the Task Force on Genetic Testing (TFGT) and the US Department of Health and Human Services' National Action Plan on Breast Cancer (NAPBC) Hereditary Susceptibility Working Group. Both had suggested that much more research needed to be conducted with regard to the clinical, psychological, and social implications of BRCA testing, and that overall, the new technology should only be provided in the context of counseling. Murphy was likely to bring these insights to her work at Oncormed.

Murphy sought to develop a service that would integrate counseling and testing and be acceptable to the medical genetics community. She voluntarily decided to follow the more stringent recommendations offered by the TFGT, by the NAPBC Hereditary Susceptibility Working Group, by ASHG, and by ACOG: OncorMed would offer BRCA testing only in the context of clinical research.

Research was not required by the terms of Stratton's license and the recommendations of these advisory groups did not carry the force of law. In principle, Murphy could have simply offered OncorMed's BRCA-testing service to anyone who wanted it, relying on Stratton's counseling guidelines and OncorMed's previous training efforts to assure that patients received appropriate counseling. The company's role could have been limited to analyzing blood samples and returning results about the client's mutation status to the health-care professional and the client. It also could have developed an approach similar to GDL's, ensuring appropriate attention to clinical care by requiring clients to purchase testing through an academic medical center. OncorMed's restriction of testing to the research context demonstrated concern about the psychosocial dimensions of testing and an interest in building a testing system that was in keeping with the norms and priorities that had been established by the medical genetics community, and it also certified this commitment because the vast majority of clinical research protocols had to be approved by an Institutional Review Board. According to American law, all research protocols conducted by investigators at an institution that receives federal grant monies—e.g., an academic medical center—must be approved by an IRB, an ethics board made up of physicians, scientists, bioethicists, and representatives of the local community.[43] An IRB examines all research protocols conducted at an institution in order to ensure that they are ethically sound and scientifically valid. As a private company receiving no federal funds, however, OncorMed was under no legal obligation to require IRB approval of its research protocols, but such approval could implicitly certify the company's commitment to excellent and appropriate health care.[44]

Of course, OncorMed was also a private company that needed to turn a profit in order to please its partners and stockholders. How could it reconcile its commitment to limiting testing services to high-risk women within clinical research protocols and its need to generate revenue and produce profits? OncorMed decided to create its own IRB-approved research protocols. According to its 1995 annual report: "Recently, we initiated our own IRB-approved national protocols for hereditary breast, ovarian, and colon cancers and familial melanoma, allowing us to broaden access to these services without compromising our high medical standards or commitment to patient protection."[45]

Seeking IRB approval for its testing system would allow Oncormed to maintain a balance between its commercial objectives and commitment to excellent care. It could increase its potential market by including individuals who did not have easy access to research protocols at an academic medical center, while also empowering a governing body which could certify that the company operated in the best interests of the users of its testing system. Patricia Murphy described OncorMed's IRB at a Senate hearing on genetic testing by noting that it allowed the company to limit testing to high-risk individuals and ensure that they received appropriate counseling: "OncorMed's protocols are designed to ensure that our susceptibility testing is provided only to people who are at high risk. Patients must be informed about the risks and limitations of the services as well as the benefits. Our protocols require that patients receive genetic counseling both before they are tested and again when the results are known."[46] In fact, as the 1990s progressed, more companies involved in research and development in the health-care sector took OncorMed's lead and began to develop IRBs or hire bioethics advisors to demonstrate their commitment to ethical standards in the development of innovation in health care.[47]

Unlike GDL's system, where IRB approval was only a factor for the fraction of its clients who participated in research protocols, IRBs (albeit multiple ones) would regulate all users of OncorMed's testing system. These regulations covered interactions between principal investigators and clients being tested at academic medical centers, as well as engagements between OncorMed, health-care professionals, and research subjects involved in the company's research protocols.

Overall, Oncormed's system was built by combining multiple, somewhat contradictory, elements of the histories of genetic medicine and health care in the United States. It clearly had commercial objectives, demonstrated by its patenting and licensing of the BRCA genes and development of a commercial testing service. Its architecture, however, also incorporated many of the recommendations that had been articulated by the biomedical and patient advocacy communities as well as advisory committees, such as the provision of testing within research and the integration of testing and counseling.

A Client's Journey through OncorMed's System

Oncormed did not advertise its service to the public. Clients were referred to the company's system through their primary-care physicians or a specialized genetics clinic. Only those defined as "high risk" (according to personal and family history) and enrolled in a research protocol (usually organized by an academic medical center or OncorMed's main facility) could have access to Oncormed's laboratory analysis. The definition of high risk, however, was not standardized and varied depending on the protocol.

After a health-care professional helped her client enroll in a research protocol that would provide access to laboratory analysis, she counseled her. Unlike GDL, which left the details of the counseling interaction up to the health-care professional, OncorMed exerted additional oversight over this counseling process. The company not only provided health-care professionals with the Stratton guidelines but also collected affidavits from them certifying that they had covered the topics it outlined, which included the benefits and risks of testing for the client and her family.[48] If the patient consented to laboratory analysis after the counseling session, her blood was drawn and sent to OncorMed, along with payment, medical and family-history information, the counseling affidavit, and documentation of informed consent.

When OncorMed received these materials, it began a laboratory analysis that differed significantly from GDL's. While GDL was experimenting with a new method of DNA analysis and examining the full sequences of both BRCA genes, OncorMed used a step-by-step approach that focused on finding a gene mutation in high-risk families. First, the laboratory searched for well-known, frequently occurring BRCA gene mutations. If the laboratory's search found no mutations, it then conducted protein truncation testing, which was said to be 80 percent sensitive,[49] for unknown mutations in regions of the gene where mutations were likely. If the laboratory still found no mutations, it sequenced the rest of the gene. Payment for the laboratory analysis followed this step-by-step approach and was variably priced depending on the rarity of the mutation. The initial search for mutations cost $500, protein truncation testing cost $800, and the final sequencing cost $800. OncorMed, like GDL, used methods for laboratory analysis that reflected its priorities. Rather than trying to refine a technique, it focused on finding mutations efficiently. Its step-by-step approach not only saved clients money but also saved the company time, energy, and resources.

After testing, OncorMed returned the results to the health-care professional. Its involvement ended there. The health-care professional then reported the results to the client and described future options for management. However, the client using OncorMed's system, like her counterpart in GDL's system, made the ultimate decision about how to incorporate the test results into her medical management.

Defining the Roles of Participants

At this point, we can see clear differences between the roles of the participants in the four architectures I have described so far. While patient advocates suggested that they, the FDA, and health-care professionals should shape both testing and counseling, professional organizations envisioned their constituents—health-care professionals—as the primary decision-makers, and GDL allowed genetics clinics the authority to direct clinical care, OncorMed controlled both the identities of the participants and the interaction between them by providing counseling guidelines and restricting access to high-risk individuals enrolled in research protocols.

The client's identity as a research subject was also quite different in OncorMed's system. While GDL's laboratory research, for example, did not directly affect the client's clinical experience (and only a subset of GDL's subjects were enrolled in clinical research protocols), and the NBCC and ASHG did not specify how research on genetic testing for breast cancer should be conducted, OncorMed's subject had to be defined as high-risk according to the academic medical center or the company, and had a standardized counseling experience. She could, however, gain access to OncorMed's system through any health-care professional, not just those at academic medical centers (as was required by GDL).

Finally, the health-care professional in OncorMed's testing system had to accept restrictions on her authority. She had to be willing to work with OncorMed's research protocols or engage in research of her own. She also had to abide by the company's eligibility criteria and promise to discuss the stipulated topics in her counseling session. These counseling requirements not only shaped the health-care professional's clinical activities but it also defined the substance of the interaction between the health-care professional and the client.

The Genetics and IVF Institute

The Genetics and IVF Institute (GIVF), a private reproductive and genetics services clinic, offered BRCA testing using a different approach than GDL or OncorMed. It built BRCA testing as a commercial service and offered it both as an integrated genetic counseling and laboratory analysis for a single fee under one roof as well as a stand-alone laboratory service that could be purchased through any physician. GIVF's service, however, was not useful to everyone as its method of laboratory analysis searched only for the three BRCA gene mutations common among individuals of Ashkenazi Jewish descent. As such, its laboratory analysis served as an inadvertent gatekeeping mechanism that restricted access differently than GDL's or OncorMed's had.

GIVF was founded in 1984 by Joseph Schulman, a pediatrician and obstetrician who was the only student of Robert Edwards and Patrick Steptoe, the British inventors of in vitro fertilization. It originally focused on providing prenatal genetic testing, in vitro fertilization, and egg and sperm donation and retrieval. By the 1990s, however, it had also developed a large menu of genetic tests, including those for biochemical markers (e.g., alphafetoprotein) and genetic mutations (e.g., Fragile X syndrome, Canavan's Disease, Tay-Sachs Disease, and sickle-cell anemia). GIVF was the first provider of genetics and infertility services to provide both medical care and laboratory testing under one roof: it took the individual through her initial appointment and counseling about the procedure to laboratory analysis and follow-up visits.

With more than 300 employees, GIVF described itself as "the country's largest private clinic offering reproductive and genetics services."[50] It certainly had the resources to add another genetic test to its already large menu, and it was also a highly visible and well-established provider with a large clientele of women who might be interested in genetic testing for breast cancer. Company officials were also particularly keen to offer BRCA testing because inherited susceptibility to breast cancer had touched the personal life of its chief executive officer, Joseph Schulman. His wife, her mother, her grandmother, and her great-grandmother, all of Ashkenazi Jewish descent, had had the disease. His wife wanted to be tested, but could find no tests available outside the context of a research protocol.[51] In light of Schulman's wife's situation and GIVF's size and resources, it is not surprising that in April 1996 it became the first provider of BRCA testing outside the research context. Because it only tested for the three BRCA gene mutations common among the Ashkenazi Jewish population, the test required little additional infrastructure and was very easy to develop. The institute first offered the test to women affiliated with GIVF and their families. Soon, however, it expanded the service and began marketing it more widely, advertising the test in Jewish newspapers.[52]

GIVF built its main BRCA-testing system in the image of its other services, as an integrated clinical and laboratory service under one roof. However, GIVF—worried that most clients would not be able to gain access to its clinics in Virginia or Maryland—allowed access to its laboratory analysis services through any physician. Not surprisingly, scientists, health-care professionals, activists, and bioethicists condemned both of these services, as they felt that GIVF had violated an informal agreement in the genetics community not to provide testing commercially until more research had been conducted. In articles and op-ed pieces published in the *New York Times*, in the *New England Journal of Medicine*, and in the *Cancer Journal*, well-known figures such as Director Francis Collins of the National Center for Human Genome Research, the Stanford University anthropologist Barbara Koenig, the Harvard University biologist Ruth Hubbard, and the NBCC activist Mary Jo Ellis Kahn argued that it was premature to offer testing outside the context of research, particularly when numerous questions remained about the risks posed by gene mutations and the effectiveness of medical management options.[53] Jewish organizations also responded to the widespread availability of BRCA-testing services. Hadassah, a Jewish women's advocacy organization, issued a policy statement that articulated its commitment to work "with the genetic testing companies, the oncology community and others to combat fear-based and stigmatizing marketing techniques and inappropriate uses of genetic tests in the Jewish community, while reiterating that current knowledge does not provide a mandate for broad-scale testing for individuals outside of a controlled research setting."[54] Like the other testing providers, however, GIVF was under no legal obligation to heed these critics or follow the informal requests and recommendations of government, scientific, or professional organizations.

Unlike OncorMed and GDL, which had built testing services that to some extent conformed to the systems proposed by critics, GIVF openly disagreed with the contention that too much was still unknown about the BRCA genes to offer testing in a non-research setting. They argued instead that women deserved to have access to the potentially life-saving technology. In an article in the *Cancer Journal*, Joseph Schulman and another physician at GIVF, Harvey Stern, wrote:

We feel that prevention is better than cure, that early diagnosis is better than late, and that breast cancer genetic predisposition testing will facilitate prevention, early diagnosis, and improved disease management. It will thus be a powerful force in the struggle to reduce the tremendous morbidity and mortality of cancer. We also are of the opinion that because women die every day of breast cancer, it is an urgent matter to make breast cancer genetic screening available now through physicians to all interested women.[55]

This rhetoric, which suggested that women had a right to demand access to all health-care interventions that were available, has a long history in the American context. Not only is it connected to the cries of female empowerment discussed earlier in this chapter, but it is also part of a larger effort, which began in earnest during the 1960s, for all patients to take charge of their own health care. Responding to a medical community that had, for decades, been primary decisionmakers in the provision of health care, with physicians sometimes not even revealing medical diagnoses to their patients, a number of physicians and philosophers suggested that patients themselves should be allowed to decide the course of their own health care. In the 1970s and the 1980s, this grew into the bioethics movement, which places considerable importance on patient autonomy and which has played an important role in shaping the provision of biomedicine and behavior of patients today.[56] Finally, the idea that clients can demand access to care is not at all surprising within America's private medical system, where the patient is treated as and acts as a consumer operating in a health marketplace.

GIVF also justified its testing service by noting that it was only providing testing for BRCA gene mutations that had a well-demonstrated relationship with breast cancer "because of the presence of the 185delAG mutation in about 1 in 100 Jewish woman [sic], its known ability to truncate the BRCA1 protein, the high risk of breast cancer in such BRCA1 mutation carriers, and the high proportion of early-onset breast cancer patients in the Jewish population who manifest 185delAG. . . ."[57] Schulman and Stern's vigorous defense of their testing service because of its utility for Jewish women highlights an important tension in the development of contemporary genetic medicine. While genetics research and the development of genetic diagnostics and therapeutics targeting close-knit ethnic groups such as the Ashkenazim could be very helpful for the health and medical care of these populations, many critics argue that the development of genetic testing targeted toward particular ethnic groups, particularly in the absence of effective therapeutics, will lead to widespread discrimination.[58]

A Client's Journey through GIVF's System

Unlike GDL and OncorMed, GIVF directly marketed its testing system to the public, placing advertisements in newspapers targeted to the Jewish population. One such advertisement in the *Los Angeles Jewish Times* (figure 2.1) emphasized that GIVF was the first medical center to make the test generally available. It also called attention to the rapid turnaround time and the confidentiality of the test results, suggesting that it protected the consumer's right to maintain her confidentiality as she learned about her gene mutation status. Prospective clients were also likely to learn about GIVF's test through the media, which covered the launch of the service both because it was the first BRCA test offered outside the context of research and because of the controversy surrounding it. In fact, even GIVF's *Los Angeles Jewish Times* advertisement noted that the launch of the

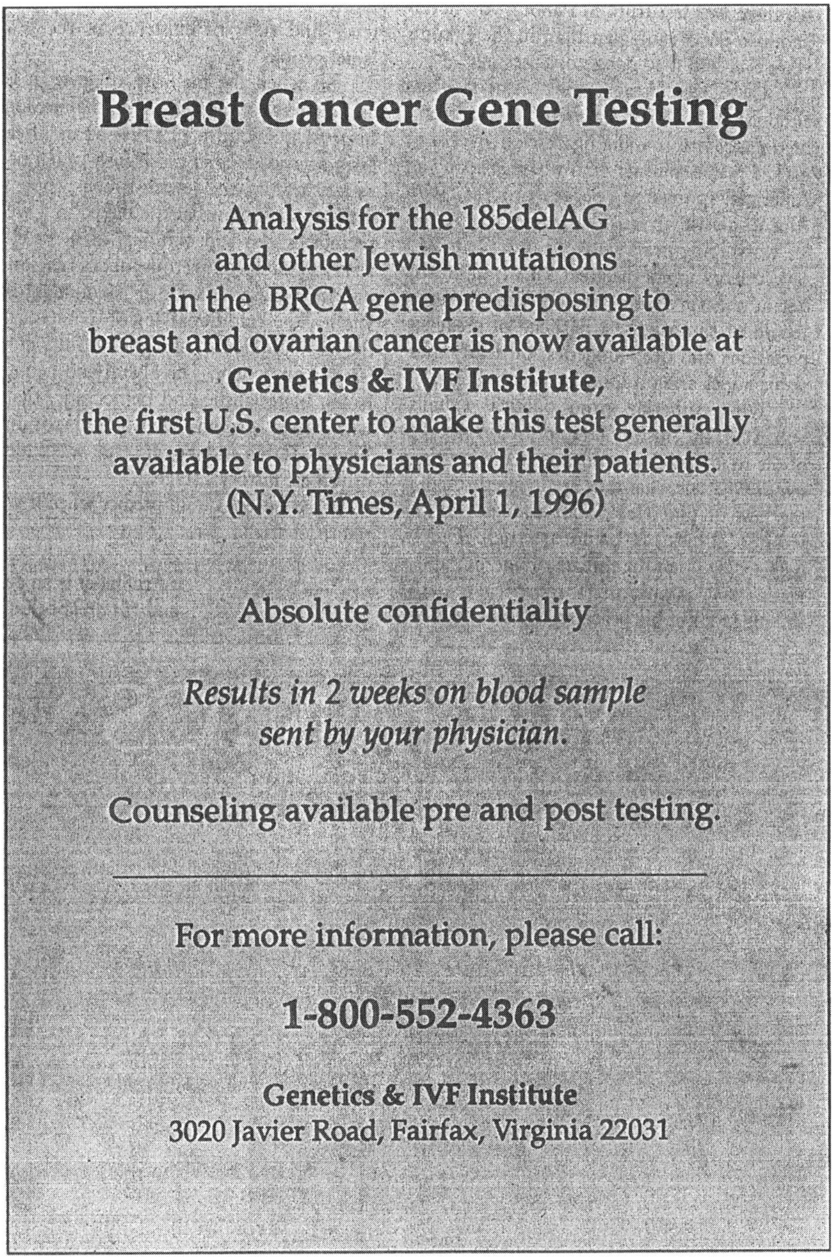

Figure 2.1
Advertisement for BRCA testing services by the Genetics and IVF Institute (*Los Angeles Jewish Times*, 1996).

service had been covered by the *New York Times* in its edition of April 1, 1996.

In order to pursue GIVF's integrated service of counseling and laboratory analysis, the client had to visit one of its clinics in either Virginia or Maryland, either on her own initiative or after the recommendation of a physician. Once she arrived at GIVF's offices, she met with a staff geneticist or genetics counselor who would gather information about family history and discuss with her the meaning of the BRCA genes, GIVF's testing system, the benefits and risks of testing, and the possible implications of testing positive for a BRCA gene mutation. Meanwhile, clients who just wanted to use GIVF's laboratory services could gain access to them through a referral from their personal physicians, and thus could bypass GIVF's counseling system. Regardless of how she got access to the system, however, if the client wanted to pursue testing, she would pay ($295 for the integrated testing and counseling service), give written certification of informed consent, and have a blood sample taken. Unlike OncorMed, which only allowed clients with extensive family histories of breast or ovarian cancer access to its system, GIVF was adamant that family history was an inaccurate predictor of the BRCA gene mutations common in the Ashkenazi Jewish population. Representatives noted: "Recent data suggest that the 185delAG mutation is highly penetrant and is likely to be associated with the same cancer risk in families with or without a history of breast cancer."[59] Thus, GIVF never rejected anyone, even if they had no family history of breast or ovarian cancer or were not of Ashkenazi Jewish descent. It is safe to assume, however, that most users of GIVF's testing system were self-selected members of the Ashkenazi Jewish population.

The blood sample was then sent to GIVF's in-house laboratory, which screened it for the three mutations common among the Ashkenazim. Because GIVF only checked for three mutations, its methods of DNA analysis were much simpler and cheaper than GDL's or OncorMed's. Once laboratory analysis was complete, GIVF either sent the result back to the independent physician who requested the test and ended its involvement or, if the client had been counseled at GIVF, a staff member conveyed the results through another counseling session. If the results were positive, the staff member and client typically discussed possible options for clinical management. If the results were negative, the client was reassured, but reminded that not only did she at least have the same breast cancer risk as the rest of the population, but also, if she had a serious family history of breast and/or ovarian cancer, she could have a BRCA gene mutation that had not been analyzed in its laboratory test. These mutations, after all, were simply the most common mutations found among the Ashkenazi Jewish population, not the only ones. As with OncorMed and GDL, GIVF's involvement in the client's health care stopped at that point. GIVF did not send test results back to the individual's primary-care physician unless the individual requested them, nor did it involve itself in the individual's post-test clinical management. Neither GIVF nor GDL and OncorMed seemed to heed the requests of patient advocates and professional organizations to ensure long-term follow-up care. Not only would it raise thorny privacy questions, but it would redefine the genetic testing system as a long term process for which they were not equipped.

Defining the Roles of System Participants
Unlike GDL, which restricted testing access to academic medical centers, and OncorMed, which used counseling guidelines, eligibility criteria based on familial risk, and research protocols to frame the roles of both health-care professionals and clients, GIVF built two testing systems that controlled its participants quite differently. The first, an integrated counseling and DNA-analysis service, managed health-care professionals and practices by keeping them under one roof. It employed health-care professionals who delivered the tests and, by providing them with both formal and informal training, also guided their counseling practices. The second, a commercial laboratory technology available through physician referral, did not interfere with the autonomy of health-care professionals or their clinical practices at all.

While the client who wanted to use either of these testing systems might be considered a consumer because she could choose which system to use and purchase DNA analysis without restrictions for a fee, her freedoms were shaped by the architecture of the testing system in one important way: the company offered only one type of DNA analysis, which searched for the three mutations common among the Ashkenazi Jewish population. While GIVF did not restrict access to this test, it is likely that this choice of DNA analysis influenced the types of clients who used its testing system.

GIVF's focus on the Ashkenazi Jewish population rested on earlier medical (and specifically genetic) interventions based on race and

ethnicity.[60] Genetics research was easier to conduct in close-knit populations with high intermarriage rates and good genealogical records, and
thus groups such as the Ashkenazim and Mormons were often the subject
of gene discovery investigations. This type of attention led to the discovery of many diseases and even specific mutations common among particular ethnic groups, and eventually the development of group-specific
screening programs for Tay-Sachs Disease and sickle-cell anemia and
"panel" tests which screened for genetic mutations in a variety of diseases
common among a particular ethnic group.[61] By providing a commercial
testing and counseling service for the three mutations common in the
Ashkenazi Jewish population, GIVF provided members of this population
with an easy opportunity to understand their genetic risk for breast and
ovarian cancer. Many critics worried, however, that genetics research
focused on particular ethnic groups and services like GIVF's would emphasize genetic differences between ethnic groups and lead societies through
a "backdoor" to eugenics.[62]

Myriad Genetics, Inc.

The fourth major testing provider was Myriad Genetics, a start-up biotechnology company based in Salt Lake City. Myriad, like GIVF, offered BRCA
testing as a commercial service, but built its architecture quite differently.
Rather than offer a package of genetic counseling and laboratory analysis,
Myriad treated BRCA testing as an ordinary medical test—the physician
ordered the test and sent payment and a blood sample to the laboratory,
and the laboratory provided the physician with the results. As a private
company with profit motives, however, Myriad also marketed its system
widely to both physicians and the public. In effect, Myriad treated BRCA
testing as a state-of-the-art product that should be available on demand to
all women.

In 1991, scientists at the University of Utah formed Myriad Genetics in
order to capitalize for gene discovery on the genealogical data that had
been collected over centuries from the large Mormon families who lived
in the state. It began by looking for one of the most highly sought after
genes yet, BRCA1, soon after Mary-Claire King localized it to chromosome
17 in 1991. Searching for such a highly anticipated gene attracted investment in the company, and by 1992 the company had entered into a 3-
year collaboration with Eli Lilly, a multinational pharmaceutical company.

Under the terms of the agreement, Lilly would provide the company with $1.8 million in return for the right to an exclusive license for diagnostic kits or therapeutic products that might result from a breast cancer gene discovery.[63] In 1994 Myriad announced it had mapped and sequenced the BRCA1 gene. By 1995, the company announced that it had mapped and sequenced the BRCA2 gene. The company applied for patents on both of these genes, adopting the same strategy as many other US gene discovery companies before it. These patents were particularly important for Myriad, as they could allow the company to control downstream innovation, including testing. Of course, because OncorMed had already applied for similar patents, both providers would be in a position to sue one another for patent infringement as soon as the patents were granted and each began testing.

While Myriad could have chosen to simply make money by licensing the patents to other companies, it decided instead to build its own BRCA-testing service. While the processes by which this decision was made are perhaps only fully understood by company executives, an outside observer might surmise that Myriad hoped to achieve several goals: to reap immediate revenue from a testing service; to develop a database of information on BRCA genes tested (and mutations found) that might eventually yield insight for therapeutic developments; and to sell or license the database to other companies. Because markets for genetic diagnostics were significantly smaller than those projected for therapeutics, it is quite likely that a genomic database would generate considerably more revenue than a BRCA-testing service.

Myriad framed itself as a private diagnostics laboratory, an organizational form that was quite familiar in the American context. It defined issues of medical practice, such as how individuals would be counseled about testing and how results would be conveyed, as being outside this scope. A company official noted, for example, how difficult it would be to assure the quality of genetic counseling:

[A]re we going to have to have people pass some sort of exam, how are we going to ensure the quality of genetic counseling? . . . And I can tell you, there were many many discussions about, let's just hire a bunch of genetic counselors and provide genetic counseling. Sort of the way, sort of the way that Genzyme [another genomics company] does it. And we did think about that very closely as well, and didn't feel that that was meeting the goals of where we wanted to go in the laboratory.[64]

Despite the recommendations of patient advocates and professional organizations that the best BRCA-testing systems would be those that were either provided in the context of counseling or through research protocols, Myriad, like GIVF, focused on offering a laboratory technology widely and quickly. In fact, in an advertisement published in the *American Journal of Human Genetics* in 1996, the company emphasized that it was taking BRCA testing out of the research setting and making it quickly available for clinical use (figure 2.2).

A Client's Journey through Myriad's System

Although it defined itself as a laboratory simply providing a DNA-analysis service, Myriad marketed its product beyond the genetics community, directly to health-care professionals and women. It advertised its service in a variety of publications, from the *American Journal of Human Genetics* to the *New York Times Magazine*. By 2002, the company had expanded its campaign to include advertisements on radio and television as well as in major women's magazines.[65] All of these advertisements provided clients with the company's toll-free number and website address, for more information about its BRCA-testing services.[66] By combining the model of an independent diagnostic laboratory with mass marketing, Myriad would likely increase its revenue from testing and, through increased testing, expand its genetic database. Not only might physicians suggest BRCA testing to their patients, as they often did with other sorts of laboratory tests, but clients might themselves initiate an inquiry.

Myriad allowed clients to gain access to its testing system through any physician, in contrast with GDL (which required clients to use academic medical centers) and with OncorMed, the NBCC, and ASHG (which restricted testing to research protocols). Myriad's client could choose to visit a genetics clinic or ask her family physician to help her gain access to BRCA testing. This diversity of health-care professionals available to clients meant that counseling could vary considerably. If a client visited a specialist at a genetics clinic, for example, she would expect to receive the benefits of training, specialization, and experience in genetics counseling. Critics have argued that primary-care physicians, by contrast, would be less likely to have formal training in genetics or genetic counseling or have the benefit of a network of colleagues (both at the institution and in professional associations) with relevant knowledge and experience.[67]

Figure 2.2
Advertisement for BRCA testing services by Myriad Genetics (*American Journal of Human Genetics*, 1996).

Myriad also did not specify eligibility criteria for its clients. While both ASCO and OncorMed suggested that only high-risk individuals be tested, Myriad did not even require health-care professionals to record a full family history of breast and ovarian cancer. While the company did ask physicians to fill out information about ancestry and clinical history (for the client and her family) on the test request form, this information did not influence the company's decision to conduct DNA analysis on the blood sample. Physicians were free to refer whomever they chose for testing; the company required only that the health-care professional send the blood sample with payment and informed consent.

Myriad offered four types of laboratory analysis: analysis of the three mutations common among the Ashkenazi Jewish population, which was also offered by GIVF (about $450); BRACAnalysis, which included full sequence analysis of both BRCA1 and BRCA2 genes (about $3,000); Rapid BRACAnalysis, which provided full sequencing of the BRCA1 and BRCA2 genes with results returned to the physician within seven days (about $4,000); and single mutation analysis (about $250), usually done after a mutation had been found in a family.[68] Myriad's method for analyzing the BRCA genes differed considerably from those conducted by GDL and OncorMed. While all of these providers checked both BRCA genes for mutations, Myriad checked for mutations as it conducted a comprehensive and standardized DNA sequence analysis of the BRCA genes while GDL targeted mutations and OncorMed targeted mutations and also sequenced parts of the genes. Myriad's laboratory methods reflected its priorities. Generating sequence data about the BRCA genes would be useful for the company's genetic database, and its focus on providing a "sophisticated" laboratory analysis—which it defined as full-sequence analysis—would reinforce the company's self-definition as merely a diagnostic laboratory.

After Myriad tested the blood sample, it sent the test results, which identified the mutation and the range of increased risk for future disease, back to the requesting health-care professional. Like GDL and OncorMed, Myriad no longer played an active role in the testing system once it returned test results.

Defining the Roles of System Participants

By defining itself as a commercial diagnostic laboratory that simply offered a DNA-analysis service, Myriad drew clear functional and temporal boundaries around the aspects of the testing system that were under its purview.

Unlike the other testing systems envisioned in the United States, it did not attempt to control directly how clients gained access to its system or how they were counseled. It also did not try to manage how health-care professionals conveyed test results to clients. It restricted its focus to providing a DNA-analysis service after a client's blood reached its laboratory. Thus, although it marketed its test directly to physicians and their clients, it did not try to get involved in their interaction.

This approach also meant that Myriad did not directly interfere with the authority of the health-care professional. Decisions about client eligibility and counseling methods before and after testing were left up to the health-care professional's discretion. In addition, Myriad did not restrict use of its system to particular health-care professionals, but instead gave access to any physician. By allowing the client to choose any physician, however, Myriad unintentionally restricted the authority of the health-care professional and discarded the systems proposed by the professional organizations. Rather than being subject to the clinical judgment of a particular specialist or the eligibility criteria of a research protocol or a genetics clinic, a client could seek the help of any physician who would facilitate her access. In fact, clients could always choose to visit another physician if one refused her access. The professional thus became a "gate opener" rather than the gatekeeper envisaged by the professional organizations and followed by most of the other American providers in one form or another.

Though both clients in GIVF's system and clients Myriad's systems can be considered consumers, because they were able to demand BRCA-testing services in exchange for payment to the test provider, the two groups of clients were defined quite differently. While GIVF's client was limited by the type of laboratory analysis available, Myriad's client was free to choose among four types of laboratory analysis and a variety of clinical interactions (including counseling at a genetics clinic and no specialized care at all). Myriad characterized this unfettered consumer choice as empowering, arguing that by choosing to purchase genetic information, clients could make their own health-care decisions. However, many breast cancer activists disputed Myriad's claim, suggesting that the information generated through commercial BRCA testing could be disempowering. In addition, in contrast to many of the other testing systems, Myriad's empowered consumer was seen as an individual, rather than part of a family or an ethnic group. While GIVF's consumers were members of one ethnic group, and Oncormed's and ASCO's clients were understood in their familial context, Myriad's client was an individual, and the answers to her questions about BRCA risk lay in her own DNA.

Thus, by the end of 1996 there was no consensus in the United States on how best to build a BRCA-testing system. Patient advocates, professional organizations, and prospective test providers disagreed about each component of the system's architecture, from the overall strategy to the methods of laboratory analysis used. They also offered different visions of testing system participants. For some, the client resembled a traditional patient, while for others, she was an empowered individual consumer of medicine. The way these groups envisioned these testing systems and their participants, however, were clearly linked to the history of genetic research, technology, and medicine in the United States and, more broadly, the structure and politics of American health care. First, the four systems represented the diversity of the genetic-testing services that were being provided in the United States: GDL's test combined research with the goals of a diagnostic laboratory, Oncormed's service wove together the priorities of a gene discovery company with the concerns of national advisory committees, patient advocates, and professional associations, GIVF's integrated clinical and laboratory system was built in the image of the private reproductive services and in-vitro fertilization clinics that were scattered across the country, while Myriad's technology was at once a simple diagnostic test and a consumer product that was in high demand. Second, the users envisioned by these technologies highlighted the contradictions of medical care in the United States. They were not simply patients engaging with the health-care system, but were often simultaneously defined as research subjects and consumers, which had additional implications for their rights and responsibilities. Meanwhile, health-care professionals had varied roles, in some cases taking on more traditional responsibilities to direct care and in others, simply facilitating consumer demand.

Britain

In many respects, the response to the BRCA gene discoveries in Britain was very similar to that in the United States. Media reports heralded the discoveries and health-care professionals, scientists, patient advocates, and prospective test providers immediately tried to influence the development of the new genetic testing technology. Unlike in the United States,

Table 2.1
Comparison of US BRCA-testing architectures.

	Breast cancer activists	Professional groups	GDL	OncorMed	GIVF	Myriad
Overall approach	NBCC: Research BCA: clinical service	ASHG, ACOG: Research Others: clinical service	Laboratory research	Clinical research	Commercial laboratory and counseling service	Commercial laboratory
Advertising	No	No	No	No	Yes	Yes
Eligibility criteria	NBCC: determined by research protocol BCA: not specified	ASHG: determined by research protocol ASCO: family history Others: Not specified	Determined by academic medical center	High-risk; Determined by research protocol	Varies: Access through any physician or GIVF's office	Access through any physician
Specialized counseling?	Yes	Yes ASHG: by genetics specialist	Yes, through academic medical center (not controlled by GDL)	Yes, using standardized guidelines	Available from GIVF staff	Variable: Decision made by clients
Laboratory method	Not specified NBCC: FDA must take active role	Not specified ASCO: labs must be regulated	CSGE	PTT and step-by-step sequencing	Mutation analysis	4 types, including full sequence analysis

Table 2.1
(continued)

	Breast cancer activists	Professional groups	GDL	OncorMed	GIVF	Myriad
Price	Not specified	Not specified	$700 for BRCA1; $1500 for BRCA2	$500 for known and frequent mutations; $800 for PTT; $800 for sequencing rest of the genes	$295 for Ashkenazi Jewish panel (three mutations common among Ashkenazim)	Rapid full-sequence analysis: ~$4000 Full sequence analysis: ~$3000 Ashkenazi Jewish panel: ~$450 1 mutation: ~$250
Post-test care	Counseling and long-term followup; reluctance to put test results in medical record	Counseling and long-term followup; reluctance to put test results in medical record	Post-test counseling through academic medical center (not controlled by GDL); reluctance to put test results in medical record	Post-test counseling, using standardized guidelines; reluctance to put test results in medical record	Varies: Post-test counseling offered by GIVF staff; reluctance to put test results in medical record	Variable: Depends on type of clinical care chosen by client; reluctance to put test results in medical record
Role of client	Right to access good medical options and be protected from bad choices	Right to be protected from bad medical choices by health care professionals	Limited right to access testing, through AMC and research protocol	Limited right to access testing (only through research protocol)	Right to demand access, limited by testing apparatus	Right to demand access, defined as empowered consumer

Role of health professional	To help protect client from bad options	Duty to educate herself and protect clients from bad choices	Duty to counsel client	Duty to counsel according to standard guidelines, participate in research and regulate client access	Duty to counsel, facilitate access to testing	Facilitate access to testing
Role of test provider	Prevent premature use, provide lab analysis and clinical care	Prevent premature use	Control lab analysis, and opportunities for access	Control eligibility, lab analysis, and clinical care	Control lab analysis and clinical care	Provide lab analysis; No involvement in clinical care
Role of government	FDA should regulate	AHA: no new regulations of labs ASCO: regulate labs; enact anti-discrimination laws	Regulate laboratory	Accepts position of government advisory committees regarding provision of testing in reserach and with counseling	Regulate laboratory	Regulate laboratory

however, it was a foregone conclusion that BRCA testing would be provided by the government's National Health Service, like all other genetic-testing services in the country. Debate then focused on how best to build a testing service that would fit with the NHS's objectives and infrastructures. How could it provide equal access across the country considering its limited resources? Should the NHS limit access to the testing system? Should the technology be controlled by regional genetics clinics, or by the central NHS administration? British groups proposed various systems to answer these challenges, and each looked quite different than its counterparts across the Atlantic.

Patient Advocates

Although patient advocates had not historically been major figures in British health-care politics, they responded to BRCA testing by following the tradition set by the British women's health movement that came before them: demanding increased access to NHS services.[69] They urged the NHS to invest in BRCA testing and broadly in genetic medicine, arguing that this new type of medical care would enhance prevention efforts and thus reduce the government's burden.

Wendy Watson, a middle-aged Derbyshire woman who had a prophylactic mastectomy in 1991 after learning of her family's extensive history of breast and ovarian cancer, spoke out immediately after the announcements of the gene discoveries. A supporter of breast cancer gene research for years, she felt that the new technology could save lives and worked hard to inform women about the test and lobby the NHS to build BRCA-testing services. In 1996, she began the Hereditary Breast Cancer Helpline, counseling thousands of people via telephone regarding their risks of breast or ovarian cancer, methods of gaining access to genetics services, and options after learning that they had tested positive for a BRCA mutation.[70] Bolstered by these interactions with individuals concerned about their BRCA risk, Watson strongly advocated the availability of BRCA-testing services on demand. In an article in *The Scotsman*, she exclaimed: "Every woman has the right to discuss her future with informed and sympathetic professionals. . . . I think it's ludicrous to say we cannot afford to fund these genetics clinics."[71] Watson argued that knowledge about her BRCA mutation status had played such an important role in her happiness and survival that other women should have the same opportunities. "Everyone should have the right to have a genetic test and take whatever action is necessary to save their lives."[72] This strong support of BRCA testing, of course, was in stark contrast to the position of American activists who characterized the new BRCA technology as of limited use and genomic information as potentially dangerous. While Watson argued that women had the right to demand genetics services, American patient advocates limited this right to medical advancements that they classified as good science, including early diagnosis of breast cancer.[73]

Watson was not alone in her support of BRCA testing. The Genetic Interest Group (GIG), which represented more than 100 organizations of individuals with genetic conditions (including cancer-research charities) and had become increasingly visible as human genetics issues were discussed by the media, Parliament, and government advisory committees, also supported increased availability of testing for all genetic conditions. In a 1995 report on the organization of UK genetics services that was disseminated to the Department of Health, regional NHS funding officials, and genetics clinics, the GIG remarked: "... it is necessary to highlight the most basic point of all—the need for an increase in the funding of the service. Backlogs are now beginning to build up in some [genetics] centres because staffing has not increased to match the rise in demand for the service. And the demand will continue to rise."[74] The report went further to recommend that services be coordinated by regional genetics clinics, be integrated with research initiatives, continue long after laboratory testing, and that laboratory testing be provided only in the context of extensive counseling.

A Client's Journey through the British Patient Advocates' System
While Watson simply emphasized that BRCA testing should be made widely available and that NHS infrastructures should be developed to support this access, the GIG provided a detailed vision of how the new technology should be provided. A client should first visit a specially trained general practitioner or secondary-care physician (e.g., an oncologist or a breast surgeon) who could gather family-history information, conduct risk assessment, and refer those at "high risk" for breast or ovarian cancer to a regional genetics clinic for further treatment. Like OncorMed in the United

States, the GIG did not define "high risk," leaving the classification up to the health-care professionals involved.

Regional genetics clinics, which were scattered across the country, would provide counseling and, if requested by the client and recommended by clinic staff, laboratory services. This integration of services resembled GIVF's system in the United States. After counseling, which included a detailed discussion of the risks and benefits of testing for the client and her family as well as the medical management options available, clinic staff sent the client's blood to their laboratory. The GIG did not specify the laboratory methods to be used, leaving this decision to the regional genetics clinic. No payment was sent with the blood sample. Whereas testing services in the United States were funded out of the client's pocket, through private insurance or research monies, the NHS would fund all services related to BRCA testing. In some cases, when the client was eligible, testing would also be reimbursed through research funds from the NHS or a medical charity.

When test results were returned to the regional genetics clinic, the health-care professional would counsel the client again about the meaning of her test results and options for the future. Advocates also suggested, like their American counterparts, that clients found to have a BRCA mutation be guaranteed access to long-term care at both the regional genetics clinics and other specialist treatment centers. To them, the testing system included not just laboratory and pre- and post-test counseling, but long term clinical interventions for the purposes of disease prevention.

Defining the Roles of System Participants

In many respects, the roles of system participants prescribed by Watson and the GIG resembled those suggested by GIVF's testing system in the United States. British patient advocates fought for provision of an integrated counseling and laboratory analysis service under one roof. Health-care professionals had a responsibility to be properly trained either to refer clients or, at the regional genetics clinic, to counsel them. The test provider (in the British case, the National Health Service) was responsible for ensuring that clinics were adequately funded, physicians were trained properly, and clients received testing and appropriate clinical care. Where GIVF and British patient advocates differed, however, was in the objectives of the system: while the NHS had public health objectives and patient advocacy groups sought to offer care to clients throughout the country, GIVF was directed simultaneously by clinical and commercial goals.

The architecture for BRCA testing proposed by British patient advocates also appears, at first glance, to be quite different than the one articulated by their American counterparts. How should we understand these differences? First, it is important to note that the two positions were actually not that far apart. Watson and the GIG presumed that BRCA testing, like other genetic tests already provided by the NHS, would be offered by specialist health-care professionals and with adequate counseling. The NBCC and BCA simply articulated the need for counseling and specialist care within an American health-care context where such services were not guaranteed and there was a real possibility that commercial providers would not insist on such services. Still, there were some important differences as the NBCC and BCA urged caution toward the new technology and Watson and the GIG pressed for increased availability. We can understand this better by considering a few key differences in the health-care politics of the two countries. First, American and British activists were operating in two very different health-care systems. American activists worried that, within a market-driven system that encouraged the rapid availability of new technologies and growth of the biotechnology industry, there were few mechanisms to regulate or even monitor development of new genetic tests. In Britain, on the other hand, BRCA testing would be provided by a trusted state-run system that was extremely popular among the citizenry. Second, the histories of patient activism were quite distinct in the two countries. In the United States, patient activists had been steadily gaining power since the women's health movements of the 1970s. Their opposition to Myriad's genetic testing system would be unlikely to seriously jeopardize the power and credibility of breast cancer activists. British activists, on the other hand, had not yet become major figures in biomedical politics. As a result, they would be much less likely to oppose practices of relatively powerful NHS clinicians. Finally, as mentioned earlier, patient activists in the two countries had traditionally been oriented toward slightly different goals. While AIDS and breast cancer activists in the United States had tried to influence the types of research conducted and medical technologies available, patient advocates in Britain had focused on simply gaining better access to NHS services, and continued to work toward this goal through their advocacy of BRCA testing.

Regional Testing Systems

In the mid 1990s, as patient advocates developed their positions toward the new technology, NHS regional genetics clinics began to build BRCA-testing systems. Many of these clinics already provided breast cancer risk-assessment services and had been involved in the research to find the breast cancer genes, and thus were well prepared to launch DNA-analysis services. All clinics modeled the new systems on the other genetic-testing services they provided—as a package of counseling and laboratory analysis connected to specialist and primary-care physicians through a hierarchical referral network. Their systems differed from the ones suggested by British patient advocates, however, by introducing geographic variation and administration of services at the regional rather than the national level.

The NHS had offered a variety of genetic-testing services at regional clinics across the country since the 1960s.[75] These clinics, which were controlled at the regional level rather than by the central NHS administration, provided counseling and offered DNA analysis through affiliated laboratories, primarily for rare disorders. NHS regional health authorities determine which genetics services would be available, and allocated a certain amount of money per year. Each regional genetics clinic, however, developed an independent strategy to offer genetics services within the regional funds that were available.[76] This administration led to regional variation in the client's access to these services, including the methods of DNA analysis used. Some clinics offered genetic testing to any client who requested the service until its annual funding allotment ran out. Others restricted testing only to those with a particularly extensive family history of disease. Thus, while all clients had to gain access to genetic testing through a regional clinic, the architectures of each system—including avenues of access and methods of laboratory analysis—varied widely.

A Client's Journey through Regional BRCA-Testing Systems

Regional clinics in Britain, unlike Myriad and GIVF in the United States, had not traditionally marketed their genetic-testing services to the public. Regions did not depart from this approach as they built BRCA-testing systems, despite the public interest in the gene discoveries. Clients learned about the new technology from their own initiative, a primary-care physician, or a specialist such as an oncologist or a surgeon, and then made an appointment with these clinics themselves or through the referral of a physician. In addition, as with previous genetic tests, all services related to BRCA testing were paid for by the NHS regional health authority.[77]

Although they all had to gain access to BRCA testing through a regional clinic, clients, once they contacted the clinic, experienced the technology quite differently by region. Waiting times to see a specialist at the genetics clinic differed, from a day to 2 years, depending upon regional demand and resources available. Access to services also varied, with each region using a different strategy to demonstrate the utility of their service to NHS authorities. Some focused on identifying BRCA mutation-positive individuals by testing only those with extensive family histories of cancer, while others interpreted the NHS's goal of providing equal access to care by offering DNA analysis on demand. Methods of laboratory analysis also demonstrated this diversity—regional laboratories housing researchers who had conducted studies on the genetics of breast cancer typically offered DNA analysis using techniques with which they were familiar, while others that had not already developed BRCA-testing methods determined laboratory protocols according to their philosophy of testing (some regions tried to stretch their NHS funding by limiting their testing analyses to those regions of the BRCA genes where mutations were most likely to be found, while others sought to do more comprehensive DNA analyses). Some laboratories used Conformation Sensitive Gel Electrophoresis (CSGE), the technique used at GDL, while others adopted Protein Truncation Testing, which had been used by OncorMed. Also, although each region had at least one diagnostic laboratory, some didn't want to build a BRCA-testing infrastructure at all, and simply outsourced these activities to other regional laboratories that were better equipped to do the DNA analysis of the BRCA genes.

None of the regional genetics clinics offered a comprehensive, full-sequence analysis of the BRCA genes like Myriad's in the United States. With its commitment to public health and equal access to care, the priorities of NHS regional genetics clinics differed considerably from the American company's. Each regional clinic received a fixed amount of funds from the NHS, and sought to maximize the clinical utility of BRCA testing by balancing investment in laboratory analysis with attention to pre- and post-test counseling. In its 1996 Best Practice guidelines for dealing with familial breast and ovarian cancer, for example, the Clinical Molecular

Genetics Society, a branch of the British Society for Human Genetics devoted to molecular genetics, addressed the diagnostic and preventative dimensions of testing. Instead of restricting its comments to laboratory practices or promoting a particular method of analyzing the BRCA genes, it emphasized the importance of risk assessment and counseling in conjunction with testing. The guidelines stated:

> Laboratories are asked to answer two types of clinical questions: diagnostic—is this familial breast cancer? and predictive—is this patient at risk of developing breast cancer? Because breast cancer is so common (lifetime risk of 1 in 8), and because the tests involved are so laborious and expensive, a strong family history must exist before diagnostic testing is undertaken. Criteria should be set (at the clinical level) for deciding which women are to be tested.[78]

While Myriad used its status as a diagnostic laboratory to distance itself from clinical care, laboratory scientists in Britain saw the clinical dimensions of the test (restricting access to clients with a strong family history) as more important than standardizing laboratory services. With genetic testing provided by the NHS, even laboratory professionals were concerned with how their activities would influence patient care.

In fact, in all of the regional BRCA-testing systems, like the other genetic-testing services developed before them, only the role of the genetics clinic was standardized. It had the authority to determine how to allocate NHS funds, how to provide access to BRCA testing, and what type of laboratory analysis to use. It also ensured that pre- and post-test counseling was available to everyone. Health-care professionals at genetics clinics discussed and developed their counseling practices through informal mechanisms such as conferences and meetings of national organizations such as the UK Cancer Family Study Group, which included medical geneticists, molecular geneticists, oncologists, genetic nurses, and genetic counselors throughout the United Kingdom who were involved in providing cancer genetics services or conducting research in the area of inherited cancer risk.[79] Before DNA analysis, these specialists in genetics provided the client with information about BRCA testing and its risks and benefits, recorded her family's history of breast and/or ovarian cancer, and counseled her about her risk. In contrast to Myriad's system, in which DNA analysis was standardized but clients' exposure to specialized counseling varied, the regional genetics clinics offered a standardized counseling experience but all other components of the testing system varied.

The client's counseling experience after DNA analysis was also fairly similar across regions. After a health-care professional at the regional genetics clinic met again with the client to discuss the results, she usually shared the results with primary-care physicians or other referring physicians in order to facilitate post-test clinical management (as the GIG had recommended). Unlike in the United States, where fear about the privacy of genetic information led test providers, health-care professionals, and clients to carefully control disclosure of test results to third parties (even primary-care physicians), British regional genetics clinics customarily shared test information with both primary-care physicians and other specialist health-care professionals in order to develop an ongoing medical management plan. In fact, one health-care professional at a regional genetics clinic noted that he purposely did not ask clients whether they wanted the results of BRCA tests to be shared with their general practitioners. He didn't want to prevent results from being returned to general practitioners. Testing, he suggested, served no purpose if it did not inform future care.[80] Long-term clinical care was considered to be part of the architecture of the testing system, and worries about the privacy of genetic information were not significant enough to alter this component of the technology. In a nation with a public health system, there was less concern about discrimination on the basis of genetic-test results.[81]

Defining the Roles of the Participants

These regional BRCA-testing systems configured the roles of participants in a very different manner than all test providers in the United States. Perhaps most importantly, the test providers at regional genetics clinics were much more tightly controlled by national and regional governmental authorities. The central NHS authority determined the territory covered by the regional genetics clinics and, to a great extent, the funding available, while the regional NHS authority determined the exact resources available to the genetics clinic.

Regional genetics clinics, however, maintained almost complete control over the allocation of resources and construction of testing systems. Thus, each had considerable influence over the roles of health-care professionals and clients within its geographic jurisdiction. This led to regional variations in how the identities of users were framed. Genetics clinics that allowed open referral, for example, controlled only their own counseling and methods of laboratory analysis. They did not control health-care professionals at the primary-care or secondary-care level by requiring them to assess eligibility and restrict referrals, and clients could take initiative to demand testing. Regional clinics that restricted access to high-risk clients not only managed their own counseling and laboratory activities but also directed all health-care professionals in a given region to refer only clients who exceeded a particular risk threshold. In such systems, both health-care professionals and clients were tightly controlled. Despite this variability, however, all clients, regardless of region, received counseling from a trained specialist at the genetics clinic.

Developing a National Strategy

By the end of 1996, when BRCA-testing services were available in most NHS regions, prominent British clinicians and public health officials began to criticize the diversity of regional testing systems and advocate adoption of a national standard for BRCA testing. Many argued that despite a history of genetics services controlled by region, these systems did not support the NHS goal of providing every individual in Britain with equal access to the health-care system.[82] They noted that the variety of testing systems led to "a lot of inequity and uneven quality," with some systems being driven solely by patient demand while others had strict eligibility criteria for testing.[83] Furthermore, these critics feared, the variety in BRCA-testing systems across the country would give the central administrators of the NHS an excuse to reduce funding for both current and future genetics services—as services grew, administrators could argue that genetics services were not provided to the British citizenry in an equitable manner.[84]

This group of health-care professionals, which was initially formed at meetings of the UK Cancer Family Study Group, proposed a national system that would provide British citizens equal access to BRCA testing across the country. While regional genetics clinics would still provide both clinical care and laboratory services, the strategy for providing these services would be standardized across the country using a system of familial risk assessment and triage. The construction of this national strategy took place in three successive stages: (1) publication of the Calman-Hine Report, which recommended that all cancer services be provided using a triage system, (2) publication of the Harper committee report, which proposed that BRCA-testing services be limited to clients defined as high risk, and (3) development of a classificatory scheme that defined low-risk, moderate-risk, and high-risk categories and recommended services for clients in each category.

In late 1994, the Chief Medical Officers of England and Wales, Kenneth Calman and Dierdre Hine, convened the Expert Advisory Group on Cancer to respond to a series of revelations in the early 1990s that linked high cancer mortality rates in Britain to poorly organized cancer care in the NHS.[85] In 1995, the group (which also included public health officials, clinicians, scientists, health economists, and a journalist) published a report titled "A Policy Framework for Commissioning Cancer Services." It recommended that the NHS ensure that "a patient, wherever he or she lives [will] be sure that the treatment and care received is of a uniformly high standard." In addition, it suggested that cancer care should be provided through a triage system, an approach which, as discussed in chapter 1, was quite familiar in the British NHS. Three levels of care would be set up, including primary-care units (general practitioners), cancer units (oncologists or breast surgeons), and specialist cancer centers (where individuals could gain access to research protocols). Each individual would be channeled according to her specific need. She would be provided with equal access to the triage system, but the type of care she received would be determined by the diagnosis of a primary-care practitioner.

Cancer genetics professionals worked immediately to capitalize on the attention paid to cancer care by demonstrating how cancer genetics was an integral component of these services. Dr. Peter Harper, head of medical genetics at the University of Wales and a member of the Welsh regional genetics clinic, spoke with the report's authors immediately after its publication and strongly encouraged them to consider the role of genetic medicine in cancer services. Genetics could be easily integrated into their framework as a specialist service, he argued, and facilitate cancer prevention.[86] Calman and Hine responded by requesting Harper to form a committee to evaluate the relationship between genetics and cancer services. With funding from the Department of Health, he gathered a committee composed of geneticists, oncologists, nurses, counselors, surgeons, a representative from the GIG, and an economist to develop recommendations to integrate cancer genetics services into the Calman-Hine framework.

The Harper committee finished its report, titled "Genetics and Cancer Services," in December 1996.[87] It integrated its recommendations with those of the Calman-Hine committee, advocating the creation of a triage system for BRCA-testing services. The first step of the system required physicians in primary-care and cancer units to gather information about a client's family history of breast and/or ovarian cancer. This responsibility would require the additional training in genetics that had already been recommended by British patient advocates and advisory committees. A client deemed "high risk" according to this family history would be referred to the regional genetics clinic, which would serve the function of the "specialist cancer center" as defined in the Calman-Hine report. Though only "high-risk" clients would be eligible for care at the regional genetics clinic, all clients could have a family history taken by a primary-care practitioner or a secondary-level specialist and receive information about their genetic risk. The committee argued that the triage system would provide a clear mechanism to ensure that testing services were provided to the small fraction of clients who needed them rather than the large population who demanded them, as only a very small proportion of the individuals interested in the new testing technology would have a mutation in one of the BRCA genes. "There is," the report stated, "a rapidly increasing demand for these services, and also for less well validated applications in lower-risk situations for common cancers. Purchasers have until now lacked information on which activities are and are not of value, and there has been no clear mechanism for commissioning services."[88] In addition, they felt that this system would justify funding from the NHS by demonstrating a rational basis for the provision of services, because it was structured in a similar manner to other accepted cancer services that used a triage system to determine provision of care.

While the Harper committee recommended gatekeeping mechanisms that would standardize access to testing across Britain, it did not attempt to control the laboratory methods used to analyze the BRCA genes. It only noted that "laboratories undertaking presymptomatic genetic testing for familial cancers should be appropriately experienced and accredited, should be closely associated with clinical services in cancer genetics and should form part of the overall cancer center specialist services."[89] Instead, the Harper committee adopted an approach similar to the regional testing systems and previous genetics services and focused on the integrated provision of counseling and testing. "Presymptomatic genetic testing," it noted, "should be regarded as a process involving not only laboratory analysis, but provision of appropriate information to those requesting testing, as well as interpretation of any result in the light of all available clinical and genetic information."[90] In contrast to Myriad's system in the United States, the Harper committee sought to standardize how clients got access to testing and the type of care they received, rather than the laboratory methods that the regional genetics clinics used.

The question of who fit into the "high risk" category, however, remained. Soon after publication of the Harper report this issue was addressed by the UK Cancer Family Study Group as well as Dr. James Mackay, an oncologist who headed the cancer genetics clinic at the University of Cambridge and had sat on the Harper committee, and Dr. Ron Zimmern, a public health specialist and director of the NHS-funded Public Health Genetics Unit. Together, they worked to define low-risk, moderate-risk, and high-risk categories for the national classification and triage strategy, assign eligibility criteria (using family-history information), and determine the laboratory-analysis, counseling, and risk-management options that would be available for each category.[91]

Like the Harper committee, these health-care professionals proposed a national standard because they felt that genetics clinics would be unable to provide equal and appropriate care if they were overburdened by inappropriate referrals. One member of the team said: "If everybody went to regional genetic services, the genetic services would be swamped. . . . So what we are saying is, that . . . the categorization between low on the one hand and moderate and the high on the other, is really a categorization of who should be managed in primary care and who should be referred on."[92] In contrast to Myriad's and GIVF's commercial testing systems, in which clients could demand access to testing, these British health-care professionals saw demand as a problem they needed to solve.

As suggested by the Harper committee, proponents of the British national standard based their risk categorization scheme on family-history information. They acknowledged, however, that little research had been done to determine the exact relationship between an individual's family history and her risk of having a BRCA mutation. As a result, Mackay, Zimmern, and their colleagues based their scheme on studies that linked an individual's family history to her overall risk of contracting breast and ovarian cancer as well as analyses of the incidence of BRCA mutations in the population.[93] At a meeting with cancer genetics professionals, Zimmern noted, "Today is not about art nor about science, but a mixture of the two. There is no good scientific evidence to guide us."[94] The standard, however, would serve two very important purposes. Not only would it provide an objective guide to mitigate demand and justify triage to patients, but it would also demonstrate to NHS administrators that genetics clinics could develop appropriate schemes to deal with demand for genetic testing. "It will be seen," they remarked at one consensus meeting, "as a demonstration project of effective demand management throughout the country."[95]

While they provided considerable detail in defining a national risk-assessment and triage scheme, however, Mackay and Zimmern followed the approaches of the earlier regional testing systems and Harper committee by not attempting to standardize the laboratory protocols that should be used for testing. Their primary concern was to standardize clinical care across the country in order to ensure that genetics clinics were not overwhelmed and that the neediest individuals would have access to testing.

A Client's Journey through the Proposed National System

In order to gain access to this national BRCA-testing system, a client visited a primary-care or secondary-care professional who provided information about the BRCA genes and breast and ovarian cancer risk and gathered information about the client's family history of breast and ovarian cancer.[96] Using this family-history information, health-care professionals classified clients into low-risk, moderate-risk, and high-risk categories according to a detailed classification scheme (table 2.2).[97] If the client was deemed "low-risk," she was reassured and turned away. If she was deemed "moderate-risk," she could choose to go to the regional genetics clinic for additional counseling and possibly have access to a mammographic screening study. If the client was categorized as "high-risk," she was offered access to the regional genetics clinic and laboratory analysis. Clients classified as "high-risk" could also have access to increased mammography and prophylactic mastectomy. This interest in developing management strategies for both moderate-risk and high-risk clients demonstrated Mackay and Zimmern's, and more broadly the NHS's, commitment to identifying all clients at increased risk of breast or ovarian cancer. Rather than focusing on identi-

Table 2.2

Proposed categories of low, moderate, and high risk in Britain. Source: R&D Office of the Anglia and Oxford NHS Executive and Unit for Public Health Genetics, report of Consensus Meeting on the Management of Women with a Family History of Breast Cancer, p. 15.

Low risk	Those whose family histories did not fall into those with high or moderate risk.	Reassured at the primary or secondary care level and turned away
Moderate risk	Three first or second degree relatives with breast or ovarian cancer diagnosed at any age on the same side of the family, or	No access to laboratory analysis of BRCA genes Offered counseling at the regional genetics clinic level
	two first or second degree relatives with breast cancer diagnosed under 60, or ovarian cancer at any age, on the same side of the family, or	Some offered access to an observational study (investigating the effects of increased surveillance on women at moderate risk.)
	one first degree female relative with breast cancer diagnosed under 40 or 1 first degree male relative with breast cancer diagnosed at any age, or	
	a first degree relative with bilateral breast cancer	Offered access to both genetic counseling and laboratory analysis at the regional genetics clinic
High risk	Breast/breast ovarian families with 4 or more relatives on the same side of the family affected at any age	
	Breast cancer (only) families with 3 affected relatives with an average age of diagnosis <40	
	Breast/ovarian cancer families with 3 affected relatives with an average age of diagnosis of breast cancer <60	
	Families with one member with both breast and ovarian cancer	

fying the small population of clients with BRCA gene mutations, as many of the American testing systems had done, proponents of the national standard were concerned with identifying and managing the *larger* population of clients *at increased risk of breast or ovarian cancer.*

Once at the regional genetics clinic, the moderate- or high-risk client typically met with a specialist in genetics and received counseling about the meaning of BRCA testing and its risks and benefits. If the high-risk client chose to pursue laboratory analysis, one of her family members who had been affected by breast or ovarian cancer had to be tested first. Mackay

and Zimmern argued that this would increase the likelihood that a mutation found in a family was linked to disease incidence and thus would
enhance the utility of the test results. If the family member consented to
laboratory analysis of her BRCA genes, the health-care professional sent
her blood to the in-house laboratory. If she tested positive for a BRCA
mutation, then the client originally interested in testing (as well as other
family members) could be tested for the same mutation. Even if no affected
relatives of the high-risk client could be tested first, however, she
would still have access to additional management options. Proponents
of the national standard argued that these clients were still at increased
risk and should thus have access to management options, but testing her
for a mutation without information about how the mutation might be
linked to disease incidence would provide little clinically useful information. The focus of the testing system was clear: identifying and managing
high-risk clients rather than simply finding BRCA mutations. This
orientation, and its consequences, will be discussed in greater detail in
chapter 4.

After laboratory analysis, the health-care professional met with the client
to present the results and discuss post-test management options. If she
tested positive, she could continue to have access to preventive services. If
a client tested negative for a BRCA mutation, however, she would lose
access to these services. Results were then furnished to the primary-care
physician and, if the physician deemed it necessary, to a specialist in order
to determine the course of care.

Defining the Roles of the Participants

The national BRCA-testing strategy framed the roles of both the client and
the health-care professional in a way that contrasted starkly with the commercial systems of Myriad and GIVF. While these US testing providers
defined health-care professionals as facilitators who could help clients gain
access to their testing services, the British national BRCA-testing system
defined health-care professionals as gatekeepers who would determine
which clients could gain access to the genetics clinic and laboratory analysis services. In this respect, the system looked more like OncorMed's, in
which health-care professionals had considerable authority in defining
the "high-risk" category. While health-care professionals in OncorMed's
system were the ultimate decisionmakers about eligibility criteria, however,
the proponents of the national standard in Britain took on this responsi-

bility. Whereas in the regional systems genetics clinics had maintained considerable authority, in the national system health-care professionals at regional genetics clinics and at the primary and secondary level simply implemented a strategy that had been developed by system proponents.

In the British national testing system, in contrast with Myriad's and GIVF's systems, the client was considered both a citizen and a patient. As a user of a government-run testing system, she was a citizen entitled to use the testing system, regardless of her ability to pay or her geographic location. She also looked more like a traditional patient who could not demand specific services but was instead subject to the risk-assessment standards of the system and the clinical judgment of the health-care professional. This client was also envisioned as part of a family, in that her access to testing was restricted not only by family history (as in Oncormed's system) but also by the availability of a family member who had suffered from breast or ovarian cancer and could be tested first.

While British patient advocates and proponents of regional and national systems all envisioned BRCA-testing services funded by the NHS, they disagreed about both the overall purpose of testing and the roles of system participants. Patient advocates, who focused on finding those with BRCA gene mutations, suggested that clients had a right to testing and health-care professionals had a duty to train themselves so they could offer specialized counseling services. While regional systems also focused on finding individuals with BRCA mutations, the roles of the health-care professional and the client varied by region. Proponents of the national standard took a different approach than patient advocates or regional genetics clinics, proposing that testing should be used to find and manage all individuals— nationwide—with an elevated risk of breast or ovarian cancer, whether or not they had a BRCA mutation. Within this system, which reflected a public-health orientation, health-care professionals took on an active gatekeeping role while clients were entitled to risk assessment and triage but could not demand counseling or laboratory analysis at the regional genetics clinic.

As each of these groups envisioned their testing systems, they adopted elements that were specific to a British approach to medical genetics and health care. First, they all worked within the existing infrastructure of the National Health Service. Second, patient advocates developed positions that incorporated their history of lobbying for increased services. Third,

Table 2.3
Comparison of British BRCA-testing architectures.

	Patient advocates	Regional systems	National strategy
Overall approach	Widely available clinical service	Regionally administered clinical service	National risk assessment and triage system, clinical service, emphasis on prevention
Advertising	No	No	No
Eligibility criteria	Not specified	Variable by region	Defined by family history
Specialized counseling?	Yes	Yes	Yes
Laboratory method	Not specified	Variable by region	Variable by region
Price	Borne by NHS	Borne by NHS	Borne by NHS
Post-test care	Post-test counseling and long-term followup	Post-test counseling, long-term followup, integrated with client's overall health care	Post-test counseling, long-term followup, integrated with client's overall health care
Role of client	Right to access testing (empowered citizen)	Right to access varies according to rules determined by regional genetics clinic	Right of equal access to health care (citizen and patient)
Role of health-care professional	Duty to be educated and counsel clients	Duty to be educated and counsel clients; Right to direct access according to regional rules	Duty to be educated and counsel clients; Right to direct access according to risk assessment and triage system
Role of NHS	Duty to ensure access to new technologies; fund services	Duty to support regional services through funding	Duty to ensure equal access to testing services nationwide; develop national clinical standards, fund services

both regional genetics clinics and national system proponents sought to reach the same goal, to offer health care to an entire nation, albeit in rather different ways. Finally, as each put together the architecture of their system, they used tools that had already proven to be acceptable in the British health-care system, such as regional competition, development of national standards, and risk assessment and triage.

Conclusion

The discoveries of the two BRCA genes gave birth to multiple testing systems in both the United States and Britain. While they varied widely, from a government-run national service that was part of preventive care to a laboratory technology marketed directly to consumers, each clearly incorporated into its technological architectures elements of toolkits that were specific to national context.

The culture of privatized medicine figured in the development of US testing systems in a variety of ways. Both Myriad and OncorMed followed in the footsteps of many start-up biotechnology companies before them by placing intellectual property rights at the center of their business strategies and designing testing technologies to maximize revenue. The market approach also shaped the emphasis on confidentiality between health-care professional and client, the characterization of the client as a consumer, and the emergence of a competitive environment with multiple, diverse providers of the new technology. Meanwhile, the government's focus on regulating laboratory activities and the focus of genomics and biotechnology companies on the immediate commercialization of their patented inventions also affected the types of services proposed. (Myriad, for example, focused on perfecting its laboratory methods rather than its clinical activities.) Finally, the rich history of patient activism in the United States provided breast cancer activists with the confidence and platform to articulate a clear vision of their ideal BRCA-testing system, which contradicted those of many scientific and medical organizations and testing providers.

In Britain, groups envisioned BRCA-testing services that built upon an existing infrastructure, using NHS-funded regional genetics clinics as tertiary-care centers that offered both specialized counseling and DNA analysis. The idea that such services would be provided outside the NHS was simply not seriously considered. Despite this use of previous NHS genetics services as models, however, there was still variability between the proposed systems. Providers interpreted NHS goals of equal access in rather different ways, with some advocating risk-assessment and triage strategies and others offering more widespread availability. There were also different opinions about whether the testing systems would best be administered at the regional or national level. Meanwhile, British patient advocates, who had not been as historically powerful as their American counterparts, sought to participate in the policy process by broadly supporting development of and access to the new technology without alienating any of the scientific or medical experts involved in these discussions.

The variety of technological architectures that were built in the United States and Britain had significant implications for the users of these tests, shaping their rights and responsibilities in different ways. Oncormed's restriction of testing to high-risk clients enrolled in research protocols, for example, meant that health-care professionals had to act as gatekeepers. These gatekeepers, however, did not have the authority to provide clinical care however they saw fit—they had to adhere to the rules of research protocols as well as counsel clients according to guidelines specified by the company. Testing system architectures also envisioned their users in ways that were quite familiar in their American and British national contexts. The clients using Myriad's and GIVF's systems, for example, both looked like quintessential empowered consumers of American twenty-first-century medicine, while Oncormed's and GDL's clients were also familiar figures occupying the sometimes uncomfortable (as we will see in more detail in chapter 3) dual identities of patients and research subjects. Meanwhile, in Britain, clients of the various testing systems were not just traditional patients subject to the authority of physicians within a national health system, but also citizens protected and accorded rights (of equal access, for example) by the National Health Service. But as we have seen, although the clients envisioned by all the British testing systems incorporated aspects of both citizen and patient identities, specific technological architectures defined the rights and responsibilities of these users in particular ways.

In both the United States and Britain, however, this plurality of testing systems and the users they envisioned would not last long.

3   Eliminating the Competition and Ensuring Success

Despite the variety of approaches to genetic testing for breast cancer initially available in the United States and in Britain, one system eventually came to dominate in each country. Myriad Genetics became the sole provider of genetic testing for breast cancer in the United States while proponents of the national standard were able to encourage its adoption across Britain. This chapter explores how these monopolies emerged, comparing how providers in each country adopted tactics and mobilized strategies that were both familiar and effective in their national context. In the United States, Myriad deployed its intellectual-property position and adopted empowerment rhetoric to become the dominant provider of testing, while proponents of the national standard in Britain emphasized the importance of equal access to health care and launched a professional training effort across the country. In the end, the architectures of two very different testing systems emerged with different social orders for users in the two countries. In the United States, the dominant user of genetic medicine became a consumer, whose right to demand services from the health-care professional was accompanied with many responsibilities. In Britain, the National Health Service offered free and equal, but limited, access to users nationwide, defining her simultaneously as a citizen and as a traditional patient.

United States

Among the multiple visions of breast cancer testing that had emerged in the United States by 1996, four services had been built. The University of Pennsylvania's Genetic Diagnostic Laboratory (GDL) and Oncormed tried to incorporate the concerns of patient advocates and professional organizations by offering testing in the context of research, the Genetics and IVF Institute (GIVF) sought to deal with worries about the premature commercialization of the technology by only analyzing the three mutations common among the Ashkenazi Jewish population, and Myriad Genetics offered a laboratory technology as a commercial product on a wide scale. However, this environment—in which multiple services existed and, to some extent, competed with one another—did not last long. Myriad Genetics, backed by its strong intellectual-property position, embarked on a campaign to drive the other providers out of business. Using a combination of threats and astute business maneuvers, it forced the other testing providers out of the market by 1999.

Both Myriad and OncorMed had applied for a number of patents and licenses on various aspects of the BRCA1 and BRCA2 gene sequences, mutations, and testing methods immediately after the gene discoveries in the mid 1990s.[1] Patents such as these have served as important currency in the biotechnology industry, particularly in the United States. They have great value in inspiring investment by suggesting that a company is vigorously pursuing innovation, and they are often considered major items of value that might encourage acquisition by a large multinational corporation. They also provide the inventor with control over how the patent is used.

By 1997, the US Patent and Trademark Office had granted Oncormed and Myriad five patents covering various aspects of the BRCA1 gene sequence. Because both companies offered testing that analyzed the BRCA genes, however, they each interfered with the other's patents. This situation could have been resolved in a number of ways: (1) the two companies could have chosen to ignore the patent interference and simply continue testing; (2) they could have negotiated a cross-license, so that each company would allow the other to continue testing under a specific set of conditions; (3) one could have simply sold the patents to the other company; or (4) they could have sued one another for patent infringement.

Oncormed struck first, choosing the last option. On August 6, 1997, when OncorMed was granted a US patent covering its BRCA1 consensus sequence, it immediately sued Myriad, arguing that Myriad was infringing on its patent by sequencing the gene through testing.[2] A few months later, on December 2, 1997, Myriad received a patent covering 47 deleterious mutations to the BRCA1 gene.[3] It then filed a patent infringement suit against

OncorMed, and added another suit against OncorMed on January 20, 1998, when it received additional patents for its version of the BRCA1 gene sequence and a predisposition diagnostic test for specific mutations.[4]

After lawsuits continued for almost a year, OncorMed decided that maintaining its BRCA-testing service was not enough of a priority to justify the costs of continuing litigation against Myriad.[5] The lawsuits were settled out of court, and in May 1998 Myriad bought OncorMed's patents and testing services, which included licenses to Mary-Claire King's patent covering BRCA1 gene markers and Mike Stratton's patent covering the BRCA2 gene, for an undisclosed sum.[6] In a statement released immediately after the litigation had concluded, chief executive officer Peter Meldrum triumphantly announced: "The litigation has been resolved and our position solidified as the predominant provider of BRCA1 and BRCA2 genetic testing. Myriad remains committed to growing BRACAnalysis and other diagnostic tests, while implementing our broader corporate strategy of utilizing proprietary genomic technologies to discover disease genes and develop therapeutics with our pharmaceutical partners."[7] Myriad had used legal and economic means to eliminate OncorMed's testing service, and as a result, the company's investigational testing regime.

This resolution was clearly important to Myriad's overall strategy. First, eliminating competitors would likely result in increased sales of its test, and therefore a bigger revenue stream. Second, strengthening its intellectual-property portfolio could be extremely valuable in attracting funds from venture capitalists and private investors. When an interviewer from the *Wall Street Journal* asked Meldrum why someone should invest in Myriad, he pointed to the company's record of proprietary gene discovery: "Based upon the rate of gene discovery . . . we have the potential of creating significant value for investors over the near term."[8] And third, by controlling all of the breast cancer testing conducted in the United States, the company could develop a comprehensive database of information about details of the BRCA genes (e.g., mutation frequency in the US population).

Armed with their patents and those it acquired from Oncormed, Myriad then sought to shut down the services of both GIVF and GDL. In early 1998, Myriad sent both of these remaining providers letters to "cease and desist," arguing that their services violated its BRCA patents by providing testing in return for payment.[9] While GIVF acquiesced quickly, GDL resisted, arguing that it was only providing testing in research protocols that were exempt from Myriad's proprietary reach. Myriad disagreed, insisting that by giving results to, and receiving payments from, health-care professionals, GDL was providing a commercial service that violated its patents. As I mentioned in the previous chapter, this conflict, over what constitutes research and what is defined simply as health care, highlights an ambiguity that is frequently controversial in American biomedicine.[10] This problem arises in many contexts. For example, principal investigators of clinical research protocols provide health care to patients as they study the safety or efficacy of a drug or medical practice. By the same token, physicians providing clinical care outside the context of research protocols often publish details of interesting medical cases in leading journals in an effort to improve broader understandings about particular symptoms or conditions. IRBs often deal with the difficulties of distinguishing between "research," which is under their purview, and "clinical care," which is not. This problematic boundary was not only negotiated by clients who had used Oncormed's short-lived testing system, but was also contested in the communications between Myriad and GDL.

In order to strengthen its position that its service was restricted to research rather than clinical care, GDL began to limit its testing service to individuals enrolled in research protocols within the National Cancer Institute's Cancer Genetics Network, a group of researchers funded by the National Institutes of Health. Myriad, unimpressed by this move, sent GDL another letter. This one contended that so long as GDL disclosed results to the patient, it was providing a commercial service (providing results in exchange for payment)—rather than conducting research—and violating Myriad's patent. In a letter to lawyers for GDL, a Myriad lawyer put the matter this way:

I appreciate your personal reassurances that Dr. Kazazian [director of GDL] has not violated these guidelines; however, it is our understanding that Dr. Kazazian did not in fact comply with these guidelines in that (1) his test results were provided to patients for the primary purpose of clinical management, and (2) he did not obtain approval from the University of Pennsylvania's IRB to perform the research, which was actually performed at another institution. Moreover, Dr. Kazazian was performing the tests on behalf of a third party customer and charging the customer a fee to conduct the test. As a result of Dr. Kazazian's activities, Myriad lost customers who chose instead to order their tests from Dr. Kazazian's laboratory. Viewed as a whole, it was evidence that Dr. Kazazian was simply engaging in commercial contract research for which he charged a fee.[11]

After a series of negotiations, Myriad forced GDL to shut down its BRCA-testing laboratory. The GDL could no longer conduct any tests (even for research purposes) that involved disclosure of results to the client. This resolution, which arose through a combination of Myriad's patent rights and legal resources and GDL's reluctance to engage in a prolonged fight with the company, allowed Myriad to control not only the provision of BRCA testing, but also the definition of research and the boundary between research and commercial services for all those who engaged in laboratory or clinical services related to the BRCA genes.

Myriad had now successfully shut down all other major providers of laboratory services that were offering tests of the BRCA genes. Any potential client who wanted to have their BRCA genes analyzed for disease-causing mutations had to use Myriad's laboratory technology, which had no specialized counseling requirements and could be accessed through any physician.[12] The company's goal now would be to develop the largest possible market for its test. This could prove to be quite a challenge, as Myriad still faced considerable criticism from many scientific and medical organizations as well as patient advocacy groups. The company had many options available. It could have tried to change the opinions of the National Breast Cancer Coalition and Breast Cancer Action, and then take advantage of their considerable influence to market the test among their constituents. This strategy, however, was unlikely to work; these groups continued to question the benefits of a genetic testing system that was provided commercially and without a requirement of specialized counseling. The company also could have chosen to alter the components of its testing system to assuage concerns, but this was unlikely as well—its testing system had been built in accordance with its commercial interests and with full awareness of its opposition. Or, it could have tried to avoid the biomedical community entirely by making the test available directly to clients "over the counter," but this reclassification would have subjected it to strict Food and Drug Administration regulations. Thus, health-care professionals would have to become, as the sociologist Michel Callon might say, an "obligatory passage point" in the referral process.[13]

The company decided to enroll users into its system by using two other strategies. First, it sought to circumvent the power of its opponents. It marketed its technology directly to the health-care professionals and clients who would use it, rather than trying simply to change the minds

of the representatives who had expressed reservations. Second, it characterized its technology as empowering, capitalizing on the decades-old
effort within American medicine to help patients become active decision-
makers in their own health care as well as the desire by physicians to
maintain their autonomy.

Enrolling Health-Care Professionals

Despite the reservations of the American Society of Human Genetics, the
American Society of Clinical Oncology, the American College of Obstetrics
and Gynecology, and the American Medical Association regarding its
laissez-faire approach to clinical care, Myriad began a multi-faceted promotional effort to inform health-care professionals across the country
about its testing service. By contacting them directly, the company hoped
that physicians and other health-care professionals would make their own
decisions about testing rather than rely on the opinions of the professional
organizations to which they belonged. Myriad combined traditional
marketing with other initiatives that would educate health-care professionals while selling them their services. In these efforts, the company
publicized its service to health-care professionals of all types, regardless
of their specialty. It partnered, for example, with LabCorp, a medical
sales and marketing company which had access to more than 200,000
primary-care physicians throughout the United States.[14] Indeed, as was
discussed in chapter 1, although many professional organizations had
urged that clients of BRCA testing be counseled by trained specialists, the
Food and Drug Administration required only that genetic tests be prescribed by a physician—it did not specify the physician's area of
expertise.

One of the company's earliest promotional efforts was the distribution
of a Professional Education Program to medical institutions across the
country. Myriad suggested that institutional leaders in academic departments or private clinics use the program, which included slides and hand-
outs, to educate health-care professionals about inherited susceptibility to
breast and ovarian cancer and Myriad's services. The program provided
information about the BRCA genes, penetrance (relationship between gene
mutation and disease incidence) of BRCA gene mutations, and the benefits of testing. However, it included little information about the type
of counseling needed and made no mention of the specialized services

available at a genetics clinic. Its intention was clear: as public awareness of the BRCA genes and the possibility of testing increased, health-care professionals of all specialties needed to be able to advise clients about their inherited breast cancer risk and direct them to Myriad's BRCA-testing service. Myriad articulated this reasoning in the introduction to this document: "The proliferation of publications in the professional and popular press has challenged the health-care community to provide their patients with up-to-date information about identification and management of hereditary cancer risk. It is therefore essential that individuals such as you who have knowledge and experience in this area continue to share information with their colleagues."[15] By informing health-care professionals about breast cancer genetics and BRCA testing as more and more people became curious about their BRCA risk, Myriad could increase uptake of its technology.

Myriad also worked to make knowledge about BRCA risk and testing part of every physician's expertise by providing the AMA with an unrestricted educational grant to develop a Continuing Medical Education (CME) module on genetic testing for breast and ovarian cancer.[16] Physicians could choose among this or many other modules in order to maintain their accreditation each year. The BRCA-testing module provided information about various aspects of the technology, including the BRCA genes themselves, BRCA analysis and testing methodologies, cancer risks associated with BRCA mutations, identification of individuals at increased risk of BRCA mutations, management strategies, and the implications of genetic testing. While the module did not refer specifically to Myriad's testing services and in fact noted that "a variety of laboratory techniques have been developed to identify BRCA mutations," only Myriad's BRACAnalysis was available when the module was published in 1999. Thus, while the module was supposed to provide impartial training about BRCA testing, it would implicitly publicize Myriad's testing service among the vast majority of physicians who had no knowledge or connection to the debate over how BRCA testing should be provided. While both the CME module and the Professional Education Program could be seen to raise conflict-of-interest issues because they simultaneously fulfilled promotional and educational purposes, this practice is common in American medicine. The sales representatives of pharmaceutical companies who visit doctor's offices, for example, often educate these health-care professionals as they market their latest drug or medical device.[17] These companies also often invite physicians to attend medical conferences in order to simultaneously promote and inform them about their latest products.

As it publicized its testing service through the CME module, its Professional Education Program, and even at medical gatherings it organized across the country, Myriad was very careful not to interfere with the authority of health-care professionals who might prescribe its testing service. As mentioned earlier, its Professional Education Program did not specify the type of counseling clients should receive or how health-care professionals should interact with clients interested in testing. Physicians could make their own decisions about whether they would follow the recommendations of scientific and medical organizations to send potential clients to specialized genetics clinics, or prescribe BRCA testing directly. This was an important move on the company's part. By distancing itself from the provision of clinical care, not requiring physicians to provide any specific services or follow any guidelines, it became much easier for physicians of all specialties to use Myriad's service.

Although it marketed its test generally to all physicians, Myriad also made a point to target specialists in cancer genetics who might be most likely to see clients curious about their genetic susceptibility to breast cancer. While members of this community, both individually and through the American Society of Human Genetics, had been particularly vocal in opposing approaches to BRCA testing that did not include extensive pre- and post-test counseling (including Myriad's service), they were possibly the company's largest market. Again, rather than contend with the representatives of professional organizations or change its testing system, Myriad chose to contact health-care professionals and researchers in cancer genetics at academic medical centers and build partnerships that would be mutually beneficial. The company tried to develop special relationships with genetics clinics by taking steps to support their institutionalization in medical centers and encourage the efforts of researchers in this field while not interfering with their roles or responsibilities. Such a system would improve the institutional stability of these clinics while possibly increasing BRCA-testing referrals to the company.

Supporting genetics clinics could turn out to be an excellent strategy. As was discussed in chapter 1, genetics clinics in the United States had historically been focused on providing care for and conducting research with individuals with rare disorders. With funding based on revenues from counseling services and grants for research protocols, these clinics often had less funding than the other departments in academic medical centers and thus less institutional support for their activities.[18] Health-care professionals in a genetics program at a major cancer center in California, for example, noted that because they seemed to deal with healthy clients rather than very sick patients, the only way for their medical colleagues to notice and respect them was to bring in a large number of clients via genetic testing.[19] Although these clinics had begun to grow, particularly in the face of discoveries of genes for common diseases, they still operated with relatively small budgets and brought in limited revenues and had also begun to compete with commercial clinics (including GIVF) and with stand-alone laboratories whose services could be accessed by any physician. As a result, these small genetics clinics at academic medical centers were vulnerable, particularly in an era of increasing financial pressure on academic medical centers, to provide more services with less funding.[20] One genetics clinic tried to publicize its service by creating a large billboard in the main lobby of the cancer center where it was housed, explaining the benefits of BRCA testing and encouraging potentially at-risk individuals to take advantage of the service. The billboard stated that "Early Risk Detection Can Have Substantial Benefits" and that "Knowing Family's Cancer Risk Allows You to Share Potential Life Saving Information with Relatives."[21] As these publicity efforts might have been successful in generating clients from the local community who might have visited the center for other reasons, explicitly partnering with Myriad would likely have broadened the reach of the genetics clinic.

Myriad tried to build such partnerships by developing a "Centers of Excellence" program (which was later renamed a "Referral Centers" program) to help increase the number of clients that these genetics clinics received. It chose approximately 100 genetics programs around the country to serve as sites of referral for individuals who had directly called the company curious about their BRCA risk. As genetics programs increased their clienteles through such referrals, Myriad might get additional consumption of its testing technology. According to the company, it chose Centers according to their commitment to provide genetic testing for breast cancer, and their relationship with the local community. One representative of the company explained:

[W]e look for a center that has a strong advocate, so somebody that really wants the program to succeed. We ask that they have a willingness to invest time and resources into the program, hiring additional staff if necessary. We ask them to have an interdisciplinary patient care provider approach, so they need to have a genetic counselor, nurse, physician, different specialties, perhaps psychology, we ask them to think perhaps about marketing their center, because if they are going to go through all these motions to become a really good center, it would be good to drive patients to those centers. And then we ask that they have a positive community reputation, and a good sense within the community.[22]

In order to receive a "Centers of Excellence" label, a genetics program needed to demonstrate both their clinical qualifications and their plans to market Myriad's BRACAnalysis within their communities.

According to Myriad, many medical centers approached the company about the possibility of becoming a center of excellence. It noted that clinics were particularly interested in this "opportunity for Myriad to support them in many of the activities that they do."[23] The company was careful, however, not to interfere in the activities of the genetics clinics either in terms of who had access to testing or the type of counseling individuals received.[24] In this manner, Myriad could continue to make its technology widely available while still arguing that it was placating the concerns of professional and patient organizations. It could now say that it supported the activities of genetics clinics while also ensuring widespread access to its technology. This approach also allowed the company to underscore its position that it was trying to empower consumers. Its job was to provide as many options as possible, and consumers would have the final say in how they gained access to the technology. Consumers could decide which type of physician (general practitioner or specialist geneticist) to visit, and also override the decision of one physician and visit another if, for example, they were refused access to BRCA testing.

Myriad also tried to encourage the genetics community to prescribe its testing service by demonstrating its continued commitment to research in the area of breast cancer genetics. In response to the criticism that Myriad's testing service had shut down research-based services like those at GDL, it worked to repair relations with investigators by demonstrating that its testing system would not interfere with, and perhaps might enhance, the autonomy of scientists in pursuing their individual research interests.[25]

In 2000, Myriad and the National Cancer Institute announced a Memorandum of Understanding to provide researchers within the NCI Cancer Genetics Network with Myriad's BRCA-testing services at a low rate ($1,200 for sequencing both genes). While all testing would still be conducted at

Myriad's Utah laboratory, the company agreed to return detailed information about the results. (This would include not only the interpretation information usually provided through its clinical services but also additional information about how the genes were tested and what mutations were found.[26]) Myriad approached researchers as it had approached health-care professionals. It would not interfere with research projects, so long as investigators used its testing system and paid a fee. Myriad's press release announcing the agreement demonstrated its commitment to the independence of researchers in its testing system. The president of Myriad Genetics Laboratories announced: "Thousands of women and their families will benefit from these research studies. . . . We are pleased to work with the NCI in supporting essential cancer research. Our experience in providing full DNA sequence analysis of these genes will assist researchers in their efforts to develop preventions and cures for these devastating diseases. . . ."[27] NCI officials have remarked that a great deal of time was spent negotiating the publicity for the collaboration, as Myriad hoped that this gesture to the genetics community would result in increased testing referrals.[28]

As with the use of its BRACAnalysis service for clinical purposes, Myriad would also benefit from this agreement with the NCI by increasing both revenues and information for its database. Although the Memorandum of Understanding did specify that Myriad could not reach through to assert an intellectual-property interest in any of the information generated by "the patient data stemming from Specimens," the company could record information about the genes tested (such as mutation incidence and penetrance) and sell it for large-scale data mining. It is not clear, however, how useful this agreement would be for researchers. It would probably be most useful for those investigating the psychosocial implications of BRCA testing, who did not need to do the DNA analysis themselves. Myriad's test, however, was probably more expensive than tests conducted at cost by laboratories affiliated with the genetics clinics spearheading the psychosocial research. The fate of epidemiological, molecular, and biochemical research into the genes, however, was more difficult to determine. The GDL, for example, would only be able to continue its research, trying to refine DNA analysis techniques for long and complex genes like BRCA1 and BRCA2, if it did not return results to patients. Clients were unlikely to donate samples to GDL, however, if they were not going to receive results, particularly if BRCA testing was commercially available elsewhere. These restrictions were also likely to affect the amount of epidemiological research conducted to determine the risk conferred by specific mutations. Furthermore, in the case of epidemiological research, Myriad would probably be reluctant to contribute to research that diminished the importance of its test (e.g., family members of a client found to have a mutation that conferred little or no increased risk of cancer might be less likely to take the test). Thus, it is entirely possible that the way Myriad chose to enforce its patent and build relationships with health-care professionals and researchers shaped both the health care that was provided and research that was conducted.

Enrolling the Public

Of course, convincing health-care professionals to prescribe its new genetic test would not be enough. In order to ensure the financial success of its laboratory technology, Myriad had to encourage public demand as well. While media coverage of the BRCA gene discoveries had created early excitement about the promise of testing, Myriad faced the possibility that criticism from patient advocacy groups and a short public memory would quickly depress the use of its technology. In order to encourage demand for BRCA testing, Myriad used what Bruno Latour has called an "I want what you want" strategy.[29] Appropriating the slogan "Knowledge is power" from the women's health movement, Myriad lobbied the skeptical patient-advocacy community first. At a 1996 conference attended by both scientists and activists, Myriad's Chief Scientific Officer Mark Skolnick emphasized that the company was providing an important technology that could help women make medical decisions, noting that BRCA testing should be as commonly available as a Pap smear.[30] The company's message was clear in its promotional materials for the technology as well. In an article published immediately after the discovery of BRCA1, Skolnick reiterated the power of his company's diagnostic test, noting that it would provide "knowledge that can allow [women] to make an appropriate choice about cancer detection and treatment."[31] Myriad used this "knowledge is power" rhetoric that was once solely in the domain of women's movements to expand the market for its new genetic testing technology. By convincing women that BRCA testing would provide information to help them take power in making decisions about their own health care, Myriad sought to encourage them to use its service.

Despite the company's efforts, many advocacy groups (including the National Breast Cancer Coalition and Breast Cancer Action) continued to criticize Myriad's approach and refused to attend special meetings and seminars organized by Myriad or even speak to the company's representatives. A member of one advocacy group, the National Alliance of Breast Cancer Organizations, however, eventually agreed to sit on the company's clinical advisory board in an ad hoc capacity while helping the company develop educational materials.[32] While the NABCO representative's participation allowed Myriad to include a visible patient advocate in their discussions, it did not change the opinion of other advocacy groups who continued to refuse contact with the company.

As the futility of efforts to convince most activist, scientist and health-care professional groups of the excellence of its service became clear, Myriad largely gave up and by 1998, it began to take its empowerment message directly to the public. The company marketed its testing service to the entire population of American women by placing advertisements in such diverse locations as the *New York Times Magazine*, the USAirways in-flight magazine, and Broadway's *Playbill*.[33] These advertisements emphasized the "Knowledge is Power" strategy that the company had first articulated in its initial interactions with activists. The *Times Magazine* advertisement, for example, showed a woman boldly staring straight at the camera with her arms crossed, declaring "I did something today to guard against cancer." (See figure 3.1.) By taking Myriad's test, she could do something to "guard" against the most dreaded disease among women. The ad did not, however, include the word "prevention" or the word "treatment," since the technology offered neither. What it could do was inform women about their genetic status, and their increased risk, which, the company suggested, would help them feel empowered to make decisions about disease prevention and other health-care and lifestyle choices. In essence, the company argued, its accurate and informative genetic test would facilitate a woman's empowerment.

Myriad expressed similar sentiments in educational brochures that were available to potential clients through physicians or directly from the company upon request. One such brochure stated: "Given a choice, would you rather deal with the known or the unknown?" The back of that brochure offered "Answers."[34] The company promised women both the information and the opportunity to deal with the unknown risks of breast cancer. An educational video produced by Myriad made the message clear when a woman who had undergone testing emphasized its importance in her life, simply stating "Knowledge is Power."[35] The company hoped that its empowerment

Figure 3.1
Advertisement for BRCA testing services by Myriad Genetics (*New York Times*, 1999).

message, which had failed among patient advocacy groups, would work when presented to average women concerned about their cancer risk.

Myriad's most ambitious marketing effort was a television, radio, and print advertising campaign launched in Denver and Atlanta in September 2002. In popular magazines such as *Better Homes & Gardens* and *People* and during television shows such as *Oprah*, the *Today Show*, *ER*, and the series premiere of *CSI: Miami*, the company described its product and emboldened women to "choose to do something now."[36] (See figure 3.2.) Maintaining the empowerment themes that Myriad had developed in earlier marketing efforts, the ads spotlighted a group of diverse women who could now take charge of their bodies, their health care, and their futures by being "Ready Against Cancer Now." Myriad did not articulate any of the test's potential risks, however, because it was not required to do so. The test was not classified as a drug, and therefore it was not subject to any of FDA's rules for direct-to-consumer marketing. Early market research on these advertisements suggested that the company's empowerment strategy would be effective. Among 300 women surveyed after viewing the company's ad, 85 percent said would contact their physician regarding BRCA testing and 62 percent would go so far as to switch health-care professionals in order to gain access to it.[37] This last statistic was evidence of one of Myriad's greatest triumphs. Many women felt so emboldened after viewing the company's ad that they would be willing to seek out—and even switch—health-care professionals in order to find one who would help them gain access to the test. It would be quite difficult for the caution and complicated arguments of patient advocates or scientific and medical organizations to penetrate Myriad's simple and positive empowerment message.

Myriad also tried to garner public support and emphasize its commitment to providing women with the opportunity to learn about their genomic information by developing a reimbursement structure for its expensive technology.[38] The company recognized that the costs might be prohibitive if individuals were forced to pay for the test themselves, and developed the Myriad Reimbursement Approval Program to work with insurance companies to encourage reimbursement. This was a particularly important strategy in the United States, where most people gain access to health care through private insurers.

The company made sure to publicize these efforts. When announcing an agreement with insurance company Aetna US Healthcare, Myriad

Figure 3.2
Advertisement for BRCA testing services by Myriad Genetics (*Colorado Home & Life*, 2003).

announced: "We are pleased that Aetna US Healthcare is taking this step to provide women at risk of developing cancer with access to a test that provides information that might save their lives."[39] In publicizing these agreements, Myriad emphasized not only its state-of-the-art service but also its technology's potential to empower women through genomic information. In addition, this partnership might not only expand the company's market, but could also improve its valuation. Immediately after the announcement of its agreement with Aetna, for example, Myriad's stock price went up.[40]

As the company invited reimbursement from insurance companies, it also created a space for them to play a pivotal role in shaping the client's access to testing. Within America's privatized health-care environment, insurance companies usually had considerable influence in controlling access to medical services. They decided which services they would pay for, and who was eligible for these services. This could certainly be the case for BRCA testing as well. Aetna, for example, guaranteed reimbursement for clients who met one of the following criteria: (1) two or more first-degree (e.g., mother) or second-degree relatives (e.g., aunt) relatives on the same side of the family with breast or ovarian cancer, regardless of age of diagnosis, (2) two relatives with early-onset breast or ovarian cancer, (3) a family member with a BRCA mutation, (4) breast cancer in a male patient or relative, (5) ovarian cancer at any age and breast cancer at any age, both on the same side of the family, (6) of Ashkenazi Jewish descent and a relative with breast or ovarian cancer at any age, (7) other circumstances with authorization of Aetna's medical director.[41] Other insurance companies, including Kaiser Permanente and Blue Cross and Blue Shield, developed and implemented similar reimbursement criteria.[42] Although Myriad had worked hard to get insurance companies to reimburse the use of its technology, they could also limit access to its technology. Indeed, Myriad's test was subject to the same factors that affected the rest of American medical care.

There is, however, one important caveat to the role of insurance companies in the provision of BRCA testing. Most people were reluctant to ask their insurers to reimburse the costs of genetic tests, including BRCA testing.[43] Worried about discrimination on the basis of the test results in, for example, insurance and employment, most people paid for the expensive test themselves. Thus, it was usually the cost, not the insurer, that limited a client's access to testing. While some of them offered reimbursement plans, insurance companies ultimately did not play a major role in controlling who and how BRCA testing was used.

Overall, Myriad tried to encourage acceptance of its genetic test by promising autonomy to each of the three groups that were most likely to use it—the biomedical community, potential clients, and health insurers. First, rather than dealing with the representatives of patients, health-care professionals, and scientists, it approached them directly, suggesting that they had the power to use the new technology however they wished. Second, when dealing with both health-care professionals and clients, Myriad used a familiar rhetorical strategy that focused on empowerment, suggesting that the technology could provide them with additional free-doms to take charge of their lives. The company marketed its test to health-care professionals by speaking directly to primary-care physicians, researchers, and genetic counselors, rather than dealing with their repre-sentatives. It promised primary-care physicians the tools to deal with indi-viduals curious about their risk of breast and ovarian cancer, offered specialists in genetics the opportunity to provide care however they chose and perhaps even expand the size of their clinics, and helped scientists pursue their research interests in breast cancer genetics.

When patient representatives resisted Myriad's efforts and proposed alternative approaches to testing and empowerment, Myriad circumvented the powerful advocacy community and delivered a simple message of empowerment directly to the public. It promised potential clients that its technology could provide the freedom to make independent decisions about their health care, an argument that would be quite difficult for patient advocates to refute, particularly in view of America's commercial medical environment and histories of patient activism and bioethics. The company also tried to increase use of its services by negotiating insurance reimbursements for its tests, even though such agreements could reduce access for potential clients. As it negotiated with insurance companies, it also allowed them to develop guidelines which would control who would be eligible for reimbursement. Myriad's empowerment rhetoric was likely to be quite powerful in a country that emphasized individualism and a health-care system that focused on providing excellent health care by enhancing the rights of consumers and health-care providers alike.

The company's empowerment strategy, however, had costs. As with many other areas of health-care decisionmaking in the United States, the client's autonomy came with many responsibilities. She had to educate herself about how best to get access to testing. If she chose to go to a non-specialist geneticist, she had to figure out how to properly interpret the results. In fact, one large survey has suggested that although most internists, obstetrician-gynecologists, and oncologists know very little about the genetics of breast cancer or the risks and benefits of testing, they do not hesitate to order Myriad's test for their clients.[44] Also, the diversity created by a technology based on consumer choice made it very difficult to assess how the test was actually being used and whether there was actually a "best practice" for prescription and use of BRCA testing. Such information was presumably being gathered by Myriad, but it would be understandably reluctant to publicize its findings as a determination of best practice might constrain use of its test.

Myriad's efforts seem to have been successful. As of fall 2006, despite continued calls by advocacy groups to offer BRCA testing in the form of research or in the context of counseling, more than 100,000 BRCA tests had been sold.[45]

Britain

In Britain, the various BRCA-testing systems would not be able to comfortably co-exist either. Proponents of the national strategy envisioned a standard system of risk assessment and triage that would supply clients to the regional genetics clinic; this approach required cooperation from health-care professionals at the primary- and secondary-care levels and regional genetics clinics. There would be little room for regional independence. Rather than using financial and legal power, however, British proponents of the national standard focused on convincing their peers to cooperate through a series of high-level meetings and educational efforts.

Advocates of the national strategy, which included the Public Health Genetics Unit (led by Ron Zimmern) as well as individual clinicians such as James Mackay, worked to promote it far beyond its origins in the UK Cancer Family Study Group and to convince regions across the country to eliminate their independent testing systems. Mackay, Zimmern, and their colleagues held meetings with health-care professionals at the primary-, secondary-, and tertiary- care levels, teaching them how to participate in their national strategy and convincing them to abandon their regional systems. A representative of the Public Health Genetics Unit (PHGU) explained the marketing they had done in the East Anglia region alone: "We've held meetings in each of the districts in the region, in the East Anglia region there are four districts, and we do a roadshow and talk about genetics, and we do some general things in the morning, and in the afternoon we talk about cancer and do some case studies. I'm also chairing a working party on cancer genetics in the Eastern region at the moment. . . ."[46] This was a particularly challenging task, as national system proponents had to ask health-care professionals who were already providing testing through regional systems to surrender some of their authority and follow strict guidelines for risk assessment and triage. Proponents tried to convince health-care professionals by using two strategies. First, they argued that a national BRCA-testing system would be the best way to follow the NHS goal to provide equal and appropriate care to the entire British population. Second, they argued that an objective model of risk assessment and triage would bolster the gatekeeping authority of health-care professionals in the face of considerable patient demand.

Proponents suggested that their national BRCA-testing strategy would solve the inequities of the regional systems that were then available. One proponent of the national standard commented: "There has been a lack of regional strategy and leadership. This has led to the lowering of incentive to either provide or advertise a familial service. Clinical practice across the region is very variable and poorly targeted. Staff are not trained and there are no dedicated facilities. There is a lottery approach and care is not equitable."[47] James Mackay reiterated this sentiment, arguing that massive patient demand for BRCA testing was resulting in unequal service provision across the country: "Unrealistic expectations had been created which had led to rapidly increasing demands on primary care and on breast units and regional clinical genetic centers. There was a great deal of variation in practice throughout the United Kingdom. Many referrals were inappropriate and there was no agreed strategy for dealing with the situation."[48] A strategy that focused on alleviating inequalities was likely to be particularly powerful because of the publication in 1996 of the NHS Patient's Charter which promised the British public "the right to receive health care on the basis of your clinical need, not on your ability to pay, your lifestyle or any other factor."[49] In a health system that guaranteed equal access to care, proponents hoped that health-care professionals would be particularly distressed by evidence of regional inequity of services and thus would be encouraged to adopt a national approach to BRCA testing.

Proponents of the national standard also argued that their model of risk assessment and triage would provide an objective method of directing care that would help health-care professionals maintain their authority as gate-keepers in two ways. First, in an age when patients were becoming increasingly proactive in demanding care, a standard model of risk assessment and triage could withstand criticism. One advocate of the national standard noted, "Totally unrealistic public and professional expectations have been generated. Many anxious women are seeking advice because of their family history of breast cancer. A policy for the management of those presenting with a family history of breast cancer is needed. . . ."[50] In contrast to Myriad's system in the United States, where demand for the new technology was not only desired but sought after, proponents of the British national standard saw public excitement about BRCA testing as a problem to be managed, and suggested that an objective model could provide health-care professionals with the confidence to maintain their authoritative position in relationships with clients. "The strategy outlined," said Mackay, "would allow either the primary health care team or the multidisciplinary breast care team at the district level to stratify individuals with reasonable confidence."[51] Second, the standardized model of risk assessment and triage could benefit health-care professionals and the institutionalization of genetics services by demonstrating to NHS administrators that genetic testing could be provided in keeping with NHS goals. Bruce Ponder, a University of Cambridge researcher who was heavily involved in breast cancer genetics research and who supported the national standard, said: "Implementation of this policy will make a significant contribution to the progress of clinical cancer genetics."[52] If officials in charge of NHS funding saw the development of a nationwide standard that was well integrated into the overall health-care system, he suggested, they might increase funding for BRCA and other genetics services in the future. The diverse system that Myriad had encouraged would be difficult to justify in the NHS, where the government was more likely to allocate scarce resources when best practices had been agreed upon and implemented.

As they lobbied health-care professionals across the country to use their national system, proponents did not focus on gaining the support of Wendy Watson or the Genetic Interest Group, nor did they market it to the potential clients of testing. In an environment where patient demand was viewed as a problem, this is not entirely surprising. Furthermore, as was mentioned in chapter 2, patient activists in Britain had not historically been very powerful, and the GIG and Wendy Watson were no exception. Although they had worked hard to involve themselves in the development of BRCA testing and the Harper committee had included a GIG representative in its deliberations, many health-care professionals did not know or concern themselves much with the opinions of these patient advocates—approval of the testing system by clients was not pivotal to its integration in the NHS. Whereas the American firm Myriad depended on revenues and genomic information from the clients who used its test, in Britain the costs of BRCA testing were borne by the government. Increased support for the national BRCA-testing standard by the public would only increase the workloads of already stressed health-care professionals and force scarce resources to be stretched further.

Responding to the National System

Critics expressed their hesitations about the national BRCA-testing system by questioning the logic that had shaped the national testing strategy. At a 1998 PHGU meeting that brought together British health-care professionals and other interested parties to discuss the national standard, an oncologist who ran a genetics clinic in London pointed out that the risk-assessment model was based on early data that focused on individuals with extensive family histories of breast or ovarian cancer.[53] They simply didn't know, she argued, whether individuals with less extensive family histories of breast and/or ovarian cancer had at least as high a probability of having a BRCA gene mutation, and that it would be incorrect to make such an assumption. She suggested that a national system of risk assessment and triage that prevented individuals with limited family histories of disease from being tested could lead to insufficient testing and poor health care. If, by contrast, testing were administered at the regional level, she suggested, then health-care professionals at regional genetics clinics could make testing decisions based on the regional funding available, and could test lower-risk clients if they chose to do so, rather than follow an arbitrary national standard that limited the availability of testing. As proponents of the national standard used family history to determine risk categories because they argued that BRCA mutation status was only worth knowing if it was clearly linked to disease, those who advocated BRCA testing for clients with small family histories of disease suggested that simply knowing one's mutation status—whether or not it was linked to a family history of disease—was important. Though it was possible for a client with a small family history of cancer to have a BRCA gene mutation, it was not likely that such a mutation conferred much of an increased risk. Proponents of unrestricted testing, however, argued that the information could be useful both for the client (who had a right to know if she was even at slightly increased risk) and for researchers and clinicians (who might be able to gather aggregate data about specific mutations and investigate why some mutations were more penetrant than others). This position, of course, was very similar to Myriad's in the United States.

Gareth Evans, who ran a large cancer genetics clinic in Manchester, suggested that the national standard needed to take geographic variation into account. He questioned national system proponents' assertion that there were only 20–40 high-risk families per million population, reporting that his service had already gathered 200 BRCA-mutation-positive families in the Manchester area (with a total population of 4 million) in just a few years of testing.[54] Not only did some individuals with only a few family members with breast or ovarian cancer have BRCA mutations, he noted; there might be geographic variation in mutation prevalence which would not be found if a national standard were imposed. He argued that not only was it important to simply know your BRCA risk, but also that NHS officials needed to understand, for future research and health-care purposes, whether the prevalence of BRCA mutations varied by region. Such information could not be comprehensively gathered through the national system as it had been proposed. For Evans, the best way to identify all those at risk for breast or ovarian cancer was to test widely, and then determine whether targeted strategies needed to be developed based on regional differences.

Other health-care professionals questioned the gatekeeping mechanisms defined by the national strategy by arguing that care would be compromised if strict categories of risk were maintained. Clients defined as "moderate risk" according to the national model, critics argued, had a significant risk of contracting breast or ovarian cancer and deserved more extensive care than simply access to a mammographic screening study. One physician stated that to restrict moderate-risk women to breast units was simply inappropriate. "It could be, for example, argued that family histories of moderate-risk women were more difficult to assess than those of high-risk women and that many [secondary level] breast units did not have the staff or expertise to deal with them."[55] Indeed, one could easily envision a system that provided more clinical care to those defined as moderate-risk because of their unclear status. Despite these questions, however, these participants generally agreed that "genetic testing should only be offered in the high-risk group under the guidance of a cancer geneticist in a regional genetics centre, and women categorized at low risk should generally be managed within the primary care setting."[56] As they challenged the knowledge that led to the construction of Mackay and Zimmern's risk-assessment model, all of these critics were asserting their own authority to determine how BRCA services should be provided. Most were already in charge of regional BRCA-testing systems, and were not ready to simply accept a model that would require them to relinquish their decisionmaking authority to a national gatekeeping mechanism.

Proponents of the national system tried to deal with criticism by not forcing regions to accept the exact risk categories proposed. Rather, they focused on implementing the concept of risk assessment and triage, which allowed only high-risk clients access to laboratory analysis, while moderate-risk clients had access to some specialized services. The main objective, they argued, was to create an objective national standard, not to quarrel about the details. By leaving the details of the risk categories up to individual regions, proponents hoped to improve the likelihood that the overall strategy would be implemented. A representative from the Public Health Genetics Unit was quoted as follows:

Yeah, when I say there's been general acceptance, there's been general acceptance among those who work in the field. So the basics are not being argued. The exact criteria are not agreed upon, but that's not so important. If the southwest of England want to use a different criteria than the northeast of England, then that's not a big deal. But the main thing is that there should be a strategy in the way we move forward.[57]

As they allowed for this flexibility, proponents also delegated more authority to health-care professionals at regional genetics clinics. Health-care professionals at regional genetics clinics could not only maintain some gatekeeping authority over patients, but also make decisions about what types of individuals would be classified as low, moderate, and high risk.

Overall, however, the development of a national strategy would demonstrate to the NHS that the biomedical community could agree on the standard of care, and thus deserved funding.

This strategy, to cajole and convince other players on the British policy battlefield while their American counterparts used much more adversarial techniques, should not be particularly surprising. As mentioned in the introduction, scholars of comparative politics have long observed that British policymaking is often the result of cooperative efforts while in the United States these processes are usually much more combative.[58] What we see here, though, is that these practices are not limited to what we consider to be traditional policy domains—similar dynamics emerged among scientists and health-care professionals in negotiating the appropriate shape of BRCA testing, and had significant consequences for the architecture and users of the new technology.

While national system proponents had not approached them to discuss the viability of their national system, patient advocates did register their responses. The Genetic Interest Group, who had strongly supported the availability of BRCA testing, quickly accepted the proposed national system of risk assessment and triage. A GIG representative noted: "Because it is possible to say [whether an individual is likely to have a BRCA gene mutation] by drawing up certain protocols whether you are an individual at high, medium, or low risk. And it's inappropriate to waste health-care resources, testing people for whom there are no prior indications. As it is appropriate to avoid using resources to ensure that people who fulfill the criteria do actually get that help and support. But it's a rational thing."[59] Like the supporters of the national system, the GIG hoped that its support of a standardized approach to BRCA testing would eventually lead to both a higher profile for its organization and increased attention to rare genetic disorders. A GIG representative observed: "I think people are realizing that the benefit will come by virtue of treatments for rare disorders being piggybacked onto the technology that cracks common disorders."[60] Indeed, the GIG often calmed the members of its constituency who had rare genetic disorders by arguing that its initiatives regarding common diseases such as breast cancer would improve their efforts to influence genetics policy on a larger scale.

Although Wendy Watson did not immediately express an opinion in the debate between regional and national testing systems, she vigorously advocated an increase in funding and availability of BRCA-testing services during this time. She was not only an international resource for women curious about BRCA testing through her help line, but was also repeatedly interviewed by media outlets throughout the world regarding her opinion on the new testing technology.[61] In addition, Watson offered seminars to regional purchasers of NHS services across the country, urging them to buy BRCA testing for their regions because, she argued, it had life-saving and cost-saving benefits. "I . . . explain to them about the advantages of purchasing genetic services, how much money it saves them. In my family, genetic testing saved the NHS 68,000 pounds. Simply because four of us had the genetic test before we had breast cancer or anything like that. Three of us got faulty genes, and all of us had preventive surgery at a cost of between 2 and 4,000. My sister hasn't had a preventive mastectomy, but she didn't have to. So she didn't have a 3,000 pound operation that would have been unnecessary."[62] Watson noted that she worked very hard to make the perspective of the patient clear to health-care professionals: ". . . that's been very important, to be able to empower people, give them the information, and then they do what they want, whether it is nothing, screening, preventive surgery, even radical preventive surgery. Whatever they choose, it should be their option and they should be fully supported."[63] Watson and the GIG's involvement in discussions about breast cancer risk-assessment services helped to encourage regional purchasers to fund cancer genetics clinics. Watson reported, for example, that her perspective had been received very well. Regional health authority officials seemed interested in hearing what she had to say, and she often inspired them to begin or increase funding for BRCA testing in their region. "At the end of my speech [at a regional health authority]," she recalled, "I was inundated with people who wanted to chat to me and then someone from the health authorities said that they were mortified that they haven't had 'somebody pleading the case, because people are worrying. I find it mortifying that I haven't already purchased the service, and I should be doing it with a matter of urgency.' So I'm not greeted as being, not knowing what I'm talking about."[64] Her activism on behalf of BRCA testing certainly influenced awareness and purchasing decisions among NHS officials.

As Watson and the GIG interjected themselves repeatedly in discussions about the provision of BRCA testing in Britain, they were increasingly recognized as legitimate participants. They demonstrated that they could be important contributors to the policymaking process, lobbying the NHS for more funds on behalf of the public. In fact, when the United Kingdom's National Institute for Clinical Excellence (NICE)—whose clinical guidelines were linked to NHS funding decisions—began to investigate how to appropriately identify and manage of women at risk for familial breast cancer in 2002, Wendy Watson and two other patient representatives joined the Guideline Development Group, which included Gareth Evans, James MacKay, and other experts in public health, genetics, oncology, nursing, surgery, and health economics. This inclusion of patient advocates on the NICE committee suggests that the GIG and Watson's activism may eventually be successful beyond the politics of BRCA testing, demonstrating the role that patient advocates could play in British health policymaking more generally. Only a few years earlier, journalists had been viewed as appropriate representatives of the public. Now, advocates were being asked to play that role and to take a seat at the policymaking table.

In May 2004, after about two years of deliberation, NICE issued its "Familial Breast Cancer" guidelines.[65] It proposed a national system of risk assessment and triage that mirrored the one devised by the UK Cancer Family Study Group, the Public Health Genetics Unit, and other clinicians, with three minor differences. First, the "low risk" category was replaced by a "population risk" category, in order to emphasize that individuals who were not defined as high- or moderate-risk were not necessarily at low or no risk for contracting breast cancer in their lifetimes. Instead, they were simply at the same level of risk as the rest of the population. Second, individuals at moderate risk were referred to secondary care (i.e., oncologists or breast surgeons) rather than the genetics clinic, but received the same services as in the original national system—counseling and increased surveillance. As in the originally proposed system, health-care professionals at secondary-care centers would be taught by genetics specialists how to counsel those at moderate risk. Third, thresholds for the moderate- and high-risk categories were defined more clearly. There was no room for the regional interpretation to which the original proponents of the national standard had eventually acquiesced. These rules, which health-care professionals were expected to take "fully into account when exercising their clinical judgement" but "does not . . . override the individual responsibility of health care professionals," virtually ratified the approach set forth in the original national BRCA-testing strategy.[66] The system envisioned by the original proponents of the national standard would now be promoted and maintained by the NHS in the long term, and likely to shape the development of future genetics services.

Table 3.1

Comparison of US and British BRCA-testing architectures.

	Myriad	National strategy
Overall approach	Commercial laboratory offering DNA analysis as a consumer product	National risk assessment and triage system, focused on standardizing clinical care, emphasis on prevention
Advertising	Yes	No
Eligibility criteria	Access through any physician	Defined by family history; only high-risk clients can access DNA analysis
Specialized counseling?	Variable: Decision made by clients	Yes
Laboratory method	Four types, including full sequence analysis	Not standardized; varies by region
Price	Rapid full-sequence analysis: ~$4,000 Full sequence analysis: ~$3,000 Ashkenazi Jewish panel: ~$450 One mutation: ~$250	Borne by NHS
Post-test care	Variable: depends on type of clinical care chosen by client; reluctance to put test results in medical record	Post-test counseling, long-term followup, integrated with client's overall health care
Role of client	Right to demand access, defined as empowered consumer	Right of equal access to health care, must follow recommendations of NHS and health-care professional
Role of health-care professional	Facilitate access to testing	Duty to be educated and counsel clients; must direct access according to risk assessment and triage system
Role of test provider	Provide lab analysis; No involvement in clinical care	Duty to ensure equal access to testing services nationwide; identify and manage those at elevated risk
Role of government	Regulate laboratory	Fund service, establish clinical standards, ensure equal access nationwide

Conclusion

Myriad and proponents of a national standard used very different strategies to make their BRCA-testing systems dominant in their respective countries. In the United States, Myriad took a forceful approach, eliminating
competitors by defending its intellectual-property position and then
marketing its services directly to health-care professionals and the public.
While these tactics were likely to anger the representatives of patient advocacy groups and scientific and medical organizations who had already
registered their opposition against the widespread availability of BRCA
testing as a commercial laboratory technology, the company hoped to generate public demand on such a scale that the cries of specialists and patient
representatives would be muffled. As the company marketed its technology to both health-care professionals and clients, it emphasized its empowering capacity, suggesting that it could help clients take charge of their
health care and provide health-care professionals with the freedom to offer
services however they wished.

Proponents of a national standard engaged in a much more cooperative
endeavor in Britain, taking advantage of existing bureaucratic networks to
publicize their services and encourage adoption. Although there was considerable discussion and disagreement among NHS health-care professionals over what constituted best practices and how to achieve goals of
equal access, the sense of common purpose coupled with system-wide pressure to adopt national guidelines led to convergence on the national
system of risk assessment and triage. Indeed, even patient advocates chose
to work to promote availability of the new technology rather than argue
over the details of the testing system.

The tools that these testing providers used to become dominant in their
respective countries are not at all surprising, as each adopted tactics that
were quite familiar in its national context. Myriad behaved like many other
biotechnology and pharmaceutical companies before it, aggressively protecting its intellectual property in order to maximize its revenue. Patents
were often the only real property—and thus potential profit generators—
that these companies had. Myriad also followed in the footsteps of other
companies in America's health-care market by advertising its product
widely, and built upon the existing rhetorical strategies of feminists and
women's health advocates to encourage use of its technology. Proponents

of the national standard in Britain made their system dominant by engaging in consensus-building efforts that were common not just in the health-care sector, but throughout British policymaking. The top-down validation of the system by NICE only occurred years later, long after the UK Cancer Family Study Group and the PHGU had held dozens of meetings with their colleagues across the country to convince them of the utility of their system. NICE's recommendations, however, did provide the final word for the provision and use of genetic testing because of its influence on NHS funding decisions; Such a resolution was not possible in the United States, where there was no analogous organization devoted to the regulation of clinical care. Through this process, however, patient advocates in Britain were slowly becoming more visible and active, demonstrating themselves to be important participants in health-care policymaking. Indeed, providers took advantage of the elements in their national toolkits not only in building their technology, but also as they created their dominance and ensured their success.

4 Defining a Good Health Outcome

By 1999, Myriad's service and the centralized NHS strategy had become the dominant systems of genetic testing for breast cancer in their respective countries. But once these systems had been built and stabilized, what would their implications be for clients interested in knowing if they were at elevated risk for breast and/or ovarian cancer? Would they influence existing understandings of and approaches to breast and ovarian cancer risk and disease? In both countries, the new technology would have to find its place within existing cancer prevention infrastructures that relied primarily on statistical modeling to predict risk.[1] A variety of assessment tools were already available through physicians, the media, and even the internet. Although techniques differed somewhat, each asked clients about various health and lifestyle factors—including age, weight, personal and family history of cancer, reproductive history, sexual history, cigarette and alcohol use, and amount of exercise—and compared this information with available data to create an individualized risk estimate (some provided numerical estimates, others simply classified clients into low-risk, moderate-risk, and high-risk categories, and a few offered risk percentages.) What difference would the availability of a blood test that could predict the presence or absence of a disease-causing mutation make to these assessment techniques? Would it improve existing methods of risk identification and management? In what ways?

In this chapter, I will explore how BRCA-testing systems were integrated into health care in the United States and Britain and how they influenced understandings of both risk and disease. I will argue that the significant differences in the architectures of the testing systems, as well as in how these systems were positioned within existing medical practices in the two countries, led to very different approaches to risk, disease, and treatment, and embodied different understandings of a good health outcome. In the United States, Myriad characterized its product as a technically and clinically novel technology that could reach a distinct goal: to identify BRCA-mutation-positive individuals. In Britain, by contrast, national system proponents tried to integrate the new technology into existing approaches to risk assessment and prevention and focused on identifying and managing all those at elevated risk for breast or ovarian cancer—not just those with BRCA mutations.

United States

When it first launched BRACAnalysis, Myriad immediately distanced its technology from existing efforts to prevent breast and ovarian cancer. As was discussed in previous chapters, it had characterized its testing system as technically novel, but the company defined it as clinically novel too. When Myriad described its product to health-care professionals and clinicians, it made no mention of existing risk-assessment services. Unlike Oncormed, which had tied its DNA-analysis technology to its risk-assessment services, Myriad implicitly suggested that the availability of its test superseded, or was at least irrelevant to, a statistical risk estimate. If a client felt that she had a family history of cancer, then Myriad's technology could tell her whether she was at risk. The woman featured in the company's advertisement in the *New York Times Magazine* (figure 3.1 above) was quoted as saying: "I got a blood test called BRACAnalysis. It's designed to tell your doctor if you are at significantly increased risk of getting breast or ovarian cancer. Women who test positive may have up to a 50 percent risk of getting breast cancer by age 50 and up to a 44 percent lifetime risk of getting ovarian cancer."[2] The technology was not meant to be used with other risk-assessment measures—it, alone, could "tell your doctor" whether you were at significantly increased risk of disease. Defining its technology as clinically novel would, of course, increase the market for Myriad's product. If it was unlike any other assessment method available, then it might be more likely to arouse interest in health-care professionals and clients alike. Furthermore, clients who had been unsatisfied by risk estimates might be more attracted to a DNA test that seemed to offer a yes-or-no answer. The company could also use the novelty of its technology to justify its reluctance to involve itself in clinical care. If the company had offered its technology in the context of existing risk-assessment services, its market would be constrained, at least in part, by the practices and expertise of clinical personnel. (Myriad would be forced to work only with health-care professionals who were well versed in cancer risk-assessment tools and would perhaps have to train health-care professionals themselves.) By defining its technology as novel, however, the company could justify a different kind of relationship with the clinician—one in which she simply had to withdraw blood from the client, with no complicated risk assessment necessary. As we shall see below, the company further justified the clinical novelty of its technology by articulating new risk and disease categories as well as avenues for cancer prevention.

Producing Risk Categories

As it focused on the answers that its laboratory analysis could provide, Myriad's BRCA-testing service fashioned new risk categories based on the presence or absence of a disease-causing mutation. These categories were articulated and reinforced in the test results that were returned to health-care professionals and clients as well as the company's promotional and educational materials.

The results form that was sent from the company's Salt Lake City laboratory to the referring physician described identifying information about the client, the dates on which the sample was drawn and tested, the type of analysis performed, whether or not a mutation was found, and how to interpret the results. It was accompanied by information about the technical specifications of the analysis, including the laboratory methods used, the categories of test results possible, and the logics that defined these categories. If no known disease-causing mutation was found, the form that summarized the result announced that the client was a member of the "no mutation detected" category. The form also described the meaning of the result. One such description stated:

No deleterious mutation was found in BRCA1 or BRCA2 in this individual. . . . There are other, uncommon genetic abnormalities in BRCA1 and BRCA2 that this test will not detect. This result, however, rules out the majority of abnormalities believed to be responsible for hereditary susceptibility for breast and ovarian cancer. If this individual has never had breast or ovarian cancer, it is recommended that testing an affected relative be considered to help clarify the clinical significance of this individual's negative test result.[3]

While the result made clear that a mutation had not been found in the parts of the genes where DNA analysis had been conducted, it also produced new uncertainties. The anthropologist Margaret Lock has argued that this generation of additional uncertainty will be increasingly common as we use genetic testing to calculate disease probabilities, because epidemiological information about the relationship between genetic mutations and disease incidence is developed over long periods of time while testing services are being made available immediately after genes are discovered.[4] In the case of BRCA testing, a client found to be mutation-negative might have a mutation in another gene or an area of the BRCA1 or BRCA2 gene that had not been tested, meaning that she was still at elevated genetic risk for breast and/or ovarian cancer. In technical specifications that accompanied the test results, the company acknowledged: "There may be uncommon genetic abnormalities in BRCA1 and BRCA2 that will not be detected by BRACAnalysis.[5] Also, if a client had a significant family history of breast and ovarian cancer but a mutation had not been found in her DNA, she might be at elevated risk for non-hereditary breast and ovarian cancer. The company tried to deal with both of these scenarios by suggesting that a family member who had been affected by cancer be tested afterward. If the family member did not have a BRCA gene mutation either, it was likely that family history of the disease was caused by something other than the parts of the two BRCA genes that had been tested. Of course, proponents of the British standard had dealt with this situation by *requiring* that affected family members be tested *first*. Myriad simply suggested that this might be a method of reducing uncertainty. Of course, if the company had required that an affected family member be tested first, as the British had, it would have created more impediments to testing, and its market might have gotten much smaller.

Myriad tried to discuss all the uncertainties generated by a negative test result in brochures and guides addressed to both health-care professionals and clients. In these materials, the company pointed out that health-care professionals played an important role in helping clients understand the meaning of their negative test result in the context of their personal and family history of breast and ovarian cancer. A brochure designed for clients with negative test results advised: "Be sure to discuss the significance of your results with the health-care professional who ordered the test. He or she is the best source of information about what this result means for you and your family members. Feel free to raise any questions or concerns you may have about your result."[6] Although Myriad placed primary clinical responsibility on the health-care professionals who facilitated the client's access to testing and provided them with information to guide their care, we should recall that the company was careful not to influence or regulate the activities of the health-care professional after test results were returned. Thus, how these negative results were explained to clients and understood in terms of their health care and family history was entirely dependent on the physician that was providing care.

Test results which indicated that a client was "positive for a deleterious mutation" were structured in a similar manner to negative results, including the technical specifications and a form that summarized the results. Of course, the interpretation section of these positive results differed. Here, clients were told about the location and type of mutation that had been found, and informed about their increased risk of breast and ovarian cancer. The company could not, however, provide specific risk information about the relationship between a gene mutation and increased risk of disease (except for the three mutations common among the Ashkenazi Jewish population that had been the subject of extensive study) and thus placed them all in one category: BRCA-mutation-positive individuals. For example, one test result reporting discovery of a BRCA1 mutation stated:

Although the exact risk of breast and ovarian cancer conferred by this specific mutation has not been determined, studies in high-risk families indicate that deleterious mutations in BRCA1 may confer as much as an 87% risk of breast cancer and a 44% risk of ovarian cancer by age 70 in women. Mutations in BRCA1 have been reported to confer a 20% risk of a second breast cancer within five years of the first, as well as a ten-fold increase in the risk of subsequent ovarian cancer. This mutation may also confer an increased risk of male breast cancer, as well as some other cancers.[7]

There simply had not been enough research conducted on the hundreds of other BRCA1 and BRCA2 mutations to provide risk information that was specific to each mutation, and the studies that had been conducted suggested that risk could vary widely (a lifetime risk range of 36–85 percent) with differences due to both genetic factors and interactions between gene mutations and the environment.[8] Family-history information could help to provide more refined risk estimates, but Myriad had built a testing system that did not require collection of such information. Researchers would likely have developed more refined risk estimates as they tracked families with BRCA mutations. This research became very difficult to conduct, however, because excitement over the test had encouraged its early commercialization and because Myriad's patent position had limited the type and (probably as a consequence) the amount of research that could be done. So, because it was impossible to provide an individualized risk estimate based on the gene mutation information generated by Myriad's testing system, clients who tested positive would make up a single category that was marked by an elevated risk of breast and/or ovarian cancer that ranged quite considerably.

Clients whose positive test result put them in this category were told that they had a newly identified disease, increased cancer susceptibility, which could lead to breast or ovarian cancer incidence. In an informational guide for physicians, the company stated: "When hereditary breast-ovarian cancer is due to a mutation in BRCA1 or BRCA2, increased cancer suscepti- bility is inherited as an *autosomal dominant disorder* [emphasis added]."[9] The primary symptom of this hereditary disease was the identification of a BRCA mutation. The company argued that the breast and ovarian cancers that resulted from the disorder of increased cancer susceptibility were not like any others, even those linked to family histories of these diseases. It described the distinction between hereditary and familial cancer further: "Family history for these women is . . . an important screening tool to iden- tify the possibility of hereditary breast-ovarian cancer syndrome, but only genetic testing can definitively determine whether an individual has inher- ited cancer susceptibility."[10] This approach was also articulated in AMA's continuing medical education module (which, as discussed in chapter 3, had been supported with an unrestricted grant from Myriad). The AMA module explained: "Hereditary breast cancer, in which the pattern of sus- ceptibility suggests an autosomal dominant pattern of inheritance must be distinguished from familial breast cancer, where there may be other affected relatives, but the pattern of inheritance is not as compelling and the etiol- ogy is likely to be multifactorial."[11] Even though scientists had not yet deter- mined whether these cancers were any different in their pathology from other breast and ovarian cancers that occurred in women without BRCA mutations, suggesting that clients with BRCA mutations had a unique con- dition validated the company's diagnostic technology.

By distinguishing between *hereditary* and *familial* cancers, Myriad emphasized the distinction between its system and existing services that determined familial cancer risk. It also allowed the company to downplay the importance of family-history information, which had been considered so important in the national British system, for determining testing eligibility and overall cancer risk. In fact, while the British standard had suggested that multiple family members had to be affected by breast or ovarian cancer for BRCA testing to be worthwhile, Myriad suggested that simply having one affected family member was enough. One advertisement stated: ". . . if one or more women in your family were diagnosed with breast cancer before the age of 50, or with ovarian cancer at any age, you could be at increased risk."[12] The company thus justified not only the results its DNA analysis could provide but also its attempts to create a large market by testing even those with a very limited family history of breast and ovarian cancer. It is important to note, however, that while Myriad's interest in constructing new risk and disease categories might be motivated by potential profits, Oncormed's testing system reminds us that provision by a corporate entity does not require this characterization.

One might easily assume that only positive and negative results could be generated from a test that analyzed genes for the presence of a disease-causing mutation. In fact, approximately 10–20 percent of Myriad's full-sequence tests resulted in the identification of variants that were neither clearly deleterious nor benign. These were called variants of "uncertain significance," which were usually single-base changes found primarily in non-functional areas of the gene sequence that had not been associated with incidence of breast and ovarian cancer. They could be reclassified as disease-causing mutations, however, as further research was done. A test result that reported finding such a variant noted: "Variants of this type may or may not affect the function of the protein encoded by the gene in which it is found. Therefore, the contribution of this variant to the relative risk of breast or ovarian cancer cannot be established solely from this analysis."[13] Clients who were found to have such a variant were, in essence, at risk of being at risk for hereditary breast and ovarian cancer. Myriad advised them and their health-care professionals to watch and wait, as "ongoing research may clarify the meaning of such results."[14] The company also promised free testing to members of the client's family who had been affected by breast and ovarian cancer, to determine whether they had the variant and it was therefore linked to disease incidence.

Thus, the architecture of Myriad's BRCA-testing system led to the construction of three risk categories—a group of mutation-negative individuals, a group of mutation-positive individuals afflicted with the disorder of increased cancer susceptibility, and those who were at risk of being at risk because they had a variant of uncertain significance. It also defined a new disease: inherited cancer susceptibility. The risk categories and uncertainties articulated by Myriad's testing system were a result of how the company chose to build and market its technology. As a technology based on analysis of the BRCA gene sequences, it was relatively easy to make it available immediately after the genes had been mapped and sequenced. Understanding the clinical significance of mutations and variants of uncertain significance in the genes, however, required much more investigation. If the company had conducted more research into the variants and mutations that appeared in the genes before releasing the diagnostic technology for widespread use, as some patient advocacy groups and scientific and medical organizations had suggested, it would have been able to offer more refined risk estimates for certain mutations (rather than putting them in one category) as well as reduce the number of variants of uncertain significance since detailed investigation would provide clues about the deleteriousness of these alterations. It would have also, however, delayed the widespread introduction of the technology. While critics felt that test results were less useful as specific risk information could not be provided to mutation-positive individuals and 10–20 percent of tests resulted in identification of variants of uncertain significance, Myriad argued that it had a duty to make BRCA testing available quickly to those who needed it (and obviously articulating the complexity and uncertainties of the test might diminish the market for its technology). The company defined the three risk categories so that it could offer its technology soon after it was built, without a designated period of clinical investigation, and without restrictions on the practices of health-care professionals. In fact, as was discussed in chapter 2, the company specifically promoted the rapid promotion of its technology, noting that it was bringing BRCA testing "from research protocol to clinical practice."[15]

Of course, as I have described in previous chapters, developers—and in this case, Myriad—had complete control over when a genetic test would be made available for widespread use. Although the US Food and Drug Administration has developed an elaborate infrastructure to determine whether and how drugs and medical devices should be approved for use, and where exactly the balance should be struck between the government's responsibility to protect the public from dangerous drugs and devices and its duty to make beneficial interventions available quickly, the FDA is very reluctant to engage in a similar balancing act when dealing with genetic tests.

Producing the Utility of BRCA Testing

Myriad emphasized the clinical utility of its test beyond the mere identification of novel categories of risk and disease. The company suggested that multiple clinical options were available to deal with a BRCA-mutation-positive diagnosis. It also developed a specific testing method, called Rapid BRACAnalysis, for use by a client who had recently learned of a cancer diagnosis.

In promotional and educational materials for health-care professionals and clients, the company suggested that armed with genetic information, clients at risk for "hereditary breast and ovarian cancer" could go "beyond risk" and gain access to effective medical options (figure 4.1). It presented mutation-positive individuals with three options: increased surveillance, prophylactic surgery, and chemoprevention. While such options had not yet been found to be specifically useful among mutation-positive individuals, the company noted that the interventions would be effective because they had worked among women defined as at high risk according to their family histories of breast and ovarian cancers. In suggesting the viability of these options for clients with BRCA mutations, however, the company did not emphasize the distinctions between hereditary and familial breast and ovarian cancers that had been so important in its other materials.

Clients who chose to undergo increased surveillance of their breasts and ovaries were told to perform self breast-examination and undergo frequent clinical breast examinations, annual mammographies starting at age 25, and annual or semi-annual CA-125 screenings for ovarian cancer. Such practices would not prevent cancer incidence, but would, it was hoped, catch them at a much earlier treatable stage. A potentially more effective but extreme option was prophylactic surgery, which entailed preventive removal of the breasts and/or ovaries. While such procedures would carry risks themselves—the physical risks associated with any surgery, physiological risks caused by removal of fatty and muscle tissue around the chest wall, early onset of menopause caused by ovary removal, and psychological risk instigated by breast and/or ovary removal, it might, unlike increased surveillance, prevent cancer incidence—studies had shown that surgeries had been effective in reducing cancers among high-risk women (but not necessarily those with BRCA mutations.) Myriad's Clinical Resource for Health Care Professionals stated: "Prophylactic mastectomy may be an option for some women with BRCA mutations."[16] Although the

Figure 4.1
Myriad Genetics' brochure for clients of BRCA testing.

practice had been strongly criticized by many feminists and patient advocates,[17] its availability seemed to underscore the utility of BRCA testing, in that it offered the client an opportunity to reduce her cancer risk considerably once she learned about her BRCA-mutation status. Surgery could not remove a BRCA mutation in all of the body's cells, but it could produce immediate results by removing the parts of the body that were likely to become diseased. This could be preferable to the results of increased surveillance methods which could only detect disease incidence early. In addition, Myriad argued that prophylactic mastectomy could be more effective than simple surveillance, as "no data are available regarding the efficacy of surveillance for breast cancer."[18] The suggestion of such a drastic option to deal with a BRCA mutation reaffirmed the importance of BRCA-mutation status as an indicator of breast cancer risk. The availability of a clear and effective management option justified, in essence, the utility of Myriad's testing system.

When Myriad first launched its BRCA-testing system, increased surveillance and prophylactic surgery were the only management options available. This all changed in 1998, when American researchers took the unusual step of stopping their clinical trial of tamoxifen (a drug previously known to prevent cancer growth and recurrence by blocking the effects of estrogen) 14 months early because they found that it caused a significant reduction in breast cancer incidence among high-risk women. This NCI-funded Breast Cancer Prevention Trial (BCPT) had been designed and coordinated by the National Surgical Adjuvant Breast and Bowel Project (NSABP), a group which has conducted a variety of clinical trials related to breast and colorectal cancer. It included more than 13,000 women deemed to be at high risk according to the "Gail model," one of the most widely used statistical models for predicting a woman's risk of contracting breast cancer, from 131 medical centers in the United States.[19] The model predicted risk according to a variety of factors, including age, family history of breast cancer, age at first successful pregnancy, number of breast biopsies, and age at menarche.[20]

BCPT, which was a double-blind placebo-controlled clinical trial providing half of its participants with tamoxifen and the other half a placebo, was designed to study whether tamoxifen administered for at least 5 years reduced the incidence of breast cancer. It began enrolling subjects in 1992 and reached its enrollment target by September 1997. In March 1998, however, after the first batches of enrollees had reached the 5-year mark, the Endpoint Review and Safety Monitoring Advisory Committee (a data-monitoring committee of specialists in oncology, gynecology, cardiology, biostatistics, epidemiology, and medical ethics) determined that prelimi-nary results from the subjects receiving tamoxifen clearly demonstrated the benefits of the drug (showing an almost 50 percent reduction in breast cancer incidence) and thus the trial could no longer be continued in an ethical manner. The trial had moved beyond a point of clinical equipoise, when researchers were unsure whether tamoxifen was beneficial, to a point when, researchers argued, the drug had demonstrated its utility and there-fore had to be stopped to allow those receiving the placebo with an oppor-tunity to take the drug.

News of the preventive benefits of tamoxifen were quickly reported in the media, under headlines such as "Breast Cancer Breakthrough," "Tamoxifen Lowers Risk of Breast Cancer," and "Landmark Study Hits Home Here."[21] Most articles quoted scientists, physicians, and patients who heralded the findings. A physician who had organized the participation of the University of Texas's M. D. Anderson Cancer Center in the trial said: "This is truly a historic day in the fight against cancer. . . . This is absolutely a landmark study in prevention."[22] The co-chair of the trial's Participant Advisory Board, a subject in the trial herself, echoed this physician's sen-timents: "The results are so profound I'm speechless. . . . We don't know where we are going from here but we have taken a major step to help women reduce their incidence of breast cancer."[23] Some, however, expressed caution toward the findings, and argued that the drug should only be used among very high-risk women as the study had shown that tamoxifen increased endometrial cancer risk. One cancer specialist worried: "You don't want to give someone who has a two in a thousand risk of getting breast cancer a medicine that has a five in a thousand chance of side effects."[24] The executive director of the National Women's Health Network, a women's health advocacy group, agreed: "The message we try to get across is that not everyone is at high enough risk to make this worth-while. . . . It isn't like putting iodine in your salt or chlorine in your drink-ing water. Women have already died taking this drug. It is still not clear . . . that in the long term it will save lives."[25] These health-care professionals and patients seemed to be particularly concerned with the approval and use of a potentially dangerous drug for long-term use among women who were otherwise healthy. What was the best way to balance the benefits of this drug with its risks, as well as the risks of not taking the drug at all?

In light of the BCPT finding and despite questions about side effects that were outstanding, the company producing tamoxifen, Zeneca (now AstraZeneca) Pharmaceuticals, quickly applied for FDA approval for the drug's use as a chemopreventive (it had already been approved as a chemotherapeutic). Although many patient advocacy groups continued to argue, both in the media and position papers, that tamoxifen had not been adequately studied to justify its widespread use as a chemopreventive, the FDA quickly approved its use and Myriad added the drug to its menu of management options for BRCA-mutation-positive individuals.[26] This FDA approval should be understood in the context of recent pressure from many patient activist groups (including some of the groups who opposed the approval of tamoxifen) and the pharmaceutical industry to expedite the FDA's process for approving drugs.[27] For years, these participants in the American politics of health care had charged that the FDA's approval process was too expensive and time-consuming, which had led to the development of an expedited review process during the 1980s. The rapid approval of tamoxifen as a chemopreventive was, in many respects, a triumph of this history.

Tamoxifen, which was much less drastic than prophylactic surgery and more effective than increased surveillance, seemed like the perfect solution to the disease of inherited cancer susceptibility characterized by a BRCA mutation, particularly within the late-twentieth-century biomedical paradigm of American health care.[28] It was a biochemical intervention into the body to treat a disease that lay in the DNA. It was just a pill, thus much less invasive than mastectomy. It was also thought to stop breast cancer before it started, which made it more valuable than mammography, which only identified existing breast cancers. BCPT investigators themselves suggested that tamoxifen might be useful for BRCA-mutation-positive individuals, even though BRCA-mutation status had not been part of the definition of "high risk" used in BCPT: "While . . . information [about how tamoxifen affects breast cancer risk in BRCA-mutation-positive individuals] is, as yet, unavailable, offering women who carry these mutations the option of taking tamoxifen may be considered, since doing so provides an alternative to bilateral mastectomy."[29] In fact, they planned to revisit blood samples collected from study participants to determine their BRCA-mutation status and explore how they had been affected by tamoxifen.

Myriad quickly accepted the BCPT findings and recommendations and incorporated the risk-reducing promise of tamoxifen into its marketing strategy. Almost immediately after the US tamoxifen trial was stopped, the company publicized the availability of the drug for BRCA mutation-positive women. In a 1999 guide to BRCA testing for health-care professionals, the company reported: "A study of more than 13,000 women at increased risk of breast cancer demonstrated that the use of tamoxifen for 4 to 5 years reduced the risk of breast cancer by 45%."[30] The drug seemed to legitimize the existence of Myriad's BRCA-testing system, even though it had associated risks and had been tested among women defined as high-risk who were not necessarily mutation-positive. In fact, BCPT's definition of high-risk included information about family history, which Myriad had explicitly suggested was not necessarily relevant to clients at risk for *hereditary* breast and ovarian cancer. Nevertheless, a drug to prevent breast cancer incidence seemed to make a diagnostic technology that identified at-risk clients much more useful. The company's 2001 television ad campaign publicizing its BRCA-testing service, for example, noted the availability of "effective medical options" for those who tested positive for a BRCA mutation. This sentiment was shared by many genetics specialists in the United States, who noted that discovery of the benefits of tamoxifen had allayed many fears among both health-care professionals and clients about the utility of BRCA testing.[31]

Another Useful Test

The company's commitment to demonstrating the clinical utility of its services was also clear from the development of its Rapid BRACAnalysis test, one of Myriad's laboratory analysis options described in chapter 2. For approximately $4,000, clients could get the results of a full-sequence analysis of both BRCA genes within seven days. The company, who stated that this service had been initiated at the request of surgeons, targeted this service to clients who had already been diagnosed with cancer and were about to have the tumor removed through lumpectomy. According to Gregory Crichfield, president of Myriad's laboratory: "These doctors and their patients use the valuable information from the BRACAnalysis test to assist in decision making as they face important choices following a diagnosis of breast cancer. The information provided from BRACAnalysis testing can help a woman make decisions to improve her health and quality of life."[32] Armed with information about their gene mutation status, clients with breast cancer might choose to undergo a mastectomy instead of a lumpectomy. If the client had a gene mutation, it had likely contributed to the first cancer and could possibly cause another cancer in the future. A Rapid BRACAnalysis test and subsequent mastectomy could prevent this recurrence. The company emphasized this use in a patient brochure and direct-to-consumer ads: "The worst thing about hearing you have cancer is hearing it twice."[33] This test, however, did not escape controversy. Some breast cancer activists argued that providing genetic information to a client after she had just received a diagnosis of breast cancer would compromise her ability to make measured decisions about whether she wanted to undergo a mastectomy.[34] They felt that clients could not possibly receive proper pre- and post-test counseling when they were so anxious about their health status and under such time constraints.

Overall, Myriad characterized BRCA testing as an unprecedented technology that could improve the lives and health care of women and was distinct from other risk-assessment services that had previously been available. Furthermore, it could be offered without specialist care or counseling, because the laboratory analysis itself identified specific risks and could trigger certain treatment recommendations. As the company promoted its novel technology, which identified new risks and disease and could also direct clients to new treatments, it also articulated a specific definition of a good health outcome: to identify and treat mutation-positive individuals. The company argued that this identification process was useful not only because knowing one's mutation status would provide peace of mind but also because a variety of clinical management options were available for this newly identified class of at-risk individuals. As we will see, however, this goal was quite different from the one articulated by its British counterpart.

Britain

BRCA testing was fitted into health care quite differently in Britain than in the United States. It was part of the menu of services offered by the National Health Service, and in contrast with the United States (where

Myriad had characterized it as a novel, stand-alone technology and had tried to separate its functions from previous risk-assessment services), it was seen as a tool to enhance current practices. This orientation, coupled with the distinct architecture of the national BRCA-testing system, led to very different definitions of cancer risk, disease, and treatment in Britain.

This perspective of proponents of the national strategy had been evident in the mid 1990s, when they began to envision their technology. Their goal was to build a BRCA-testing service that would be helpful to the entire population, and they realized this goal by connecting the genetic test to existing philosophies of breast cancer risk assessment. British clinicians, like their American counterparts, had traditionally used a number of assessment methods to estimate risk, incorporating family and individual reproductive history among other things. These assessments were primarily conducted at secondary and tertiary-care centers, in the offices of gynecologists, oncologists, and geneticists. Proponents of the national BRCA-testing system saw their system of risk assessment and triage as building on these existing services and infrastructure, and DNA analysis simply refined the contents of the high-risk category.

The characterization of DNA analysis as simply a tool to enhance and refine existing risk-assessment services was clear from the way the British proponents of the national standard produced and justified the risk categories. While Myriad created categories of mutation-positive clients, mutation-negative clients, and those with variants of uncertain significance to demonstrate the utility of its technology, the British national system defined risk categories that would promote the reduction of breast and ovarian cancer incidence throughout the population, whether or not the cancers were linked to a BRCA mutation. Thus, the categories were created long before DNA analysis took place, were based on family-history information, and included individuals who were at elevated risk but unlikely to have a BRCA mutation. After an early meeting that focused on developing the national BRCA-testing system, Dr. Ron Zimmern summarized this perspective: "The meeting was to be about women who presented with symptoms or signs of breast cancer or were anxious about their own risk of breast cancer because they had relatives with the disease. The main focus was to be on the management of clinical risk, and not on genetic testing or population screening."[35] In fact, clients who had been classified as high-risk but were not tested because they either decided against it or did not have a family member who had been affected by breast or ovarian cancer who could be tested first, for example, remained in the high-risk category for management purposes and had access to the same clinical options as those who tested positive for a BRCA mutation. Meanwhile, high-risk clients from families in which a mutation had been found but who tested negative for that mutation were re-classified as low-risk and no longer had access to additional management options. Clinicians argued that if the client did not have the mutation that had been found in the family (which had also been linked to cancer incidence), her risk level was simply the same as the rest of the population. Finally, the creation of the moderate-risk group most clearly embodied the British proponents' strategy—this group of clients was clearly at elevated risk for breast and ovarian cancer and thus warranted additional attention, but their risk did not warrant the use of an expensive genetic test.

The British approach, of course, differed markedly from Myriad's characterization of BRCA testing as a technology that generated a new type of information that needed to be considered separately from family history. While the American company contended that clients without family histories of either cancer might still have a BRCA mutation and be at increased risk for hereditary breast or ovarian cancer, proponents of the British system argued that family history was very important for determining the risk of contracting disease. More clients had a large family history of breast and/or ovarian cancer than had a BRCA mutation, they observed, and identifying and treating clients with family histories of breast and/or ovarian cancer would be much more useful for reducing the overall incidence of cancer in the population. Furthermore, the British argued, those with BRCA mutations but no family history of cancer probably had a mutation of low penetrance (very little increased risk of disease), and thus were not relevant to a program dedicated to managing disease risk. A physician attending one of the Public Health Genetics Unit's meetings to develop the national standard argued that focusing on familial risk would allow health-care professionals to assess and care for a larger pool of clients than looking only for those with BRCA risk:

... breast cancer could be either genetic or sporadic. Familial cancer included both sporadic cases and cases in which there was a genetic predisposition. The cases of genetic breast cancer in which the family carried a gene such as BRCA1 or BRCA2 were small. Many familial cases did not. Some would be the consequence of common environmental influences; others of genetic susceptibility, probably due to a number of as yet unidentified genetic mutations of low penetrance. Not surprisingly this complex interaction of specific genes, non-specific genetic factors and environmental influences has led to much misunderstanding and confusion.[36]

Differences in focus between clients with BRCA mutations and those with family histories of breast and ovarian cancer can be easily explained by the different strategies of the two providers. While Myriad wanted to create a large market for its test, its British counterparts sought to reduce, regardless of method, breast and ovarian cancer incidence throughout the population. This mission of a government-run public health service was understandably different from that of a for-profit company.

These understandings of risk also led to different approaches to disease in the two countries. Whereas Myriad had gone out of its way to discuss the new diseases of "inherited cancer susceptibility" and "hereditary" cancer, which only its technology could identify, British developers were less interested in focusing on the distinction between hereditary and familial cancer. Meetings devoted to building the system and integrating it into health care had not distinguished between familial and hereditary breast cancer, as the BRCA-testing services to be offered by regional genetics clinics would be integrated with existing breast cancer risk-assessment services that had been based on family-history information. Even the National Institute of Clinical Excellence guidelines noted that the goal of the national BRCA-testing system was "the classification and care of women at risk of familial breast cancer," and family history had been, for cancer as well as other genetic diseases, a clear indicator of disease risk.[37] To the NHS, the distinction between familial and hereditary cancer and the identification of a distinct disease of inherited cancer susceptibility simply wasn't important. It wasn't trying to demonstrate the utility of or create a large market for a product. Rather, it was trying to combine new tools with existing infrastructure to better prevent cancer or detect it early. The best way to do this, when dealing with altered genes that were relevant in only a small fraction of cancer cases and also conferred varied risk depending on the specific mutation, was to use family history as a guide. They believed that this strategy could provide better predictive value. If a client had a BRCA mutation but no family history of breast or ovarian cancer, they suggested, it did not matter whether or not she had a disorder of increased cancer susceptibility. It was very unlikely that she would contract cancer—and therefore it was not important for the NHS to identify or treat her.

Managing At-Risk Clients

As was mentioned in chapter 2, the British BRCA-testing system offered management options for clients classified as high-risk and for those classified as moderate-risk. Clients classified as high-risk (those with BRCA mutations as well as those classified as high-risk who did not undergo laboratory analysis) had two options available to them: prophylactic surgery and increased surveillance. System proponents argued that these measures had been proven to reduce the risk of breast and ovarian cancer. The NICE guidelines acknowledged that surgery was an effective yet risky option but might be useful for high-risk women: "Risk-reducing surgery is only appropriate for a small proportion of women with a family history of breast cancer. Women considering this option may need considerable time and support in making decisions. . . ."[38] Increased surveillance, however, was a much less risky but also possibly a less effective intervention. When describing mammography, for example, the guidelines noted that "the effectiveness of mammographic screening in younger women is controversial, though evidence of benefit in women aged 40–49 is mounting."[39] For both interventions, the NHS decided that their possible benefits seemed to outweigh the costs and possible risks.

In contrast to the United States, chemoprevention was not among the options available to high-risk clients in Britain. While the high-risk group in Britain and the BRCA-risk group in the United States did not necessarily consist of the same types of people, one could easily assume that chemoprevention would be considered a viable option for clients defined as high-risk in Britain. After all, the clinical trials that led to tamoxifen's approval in the United States investigated the effects of the drug in women who were at risk according to various measures (including family history) but did not necessarily have BRCA gene mutations. Why, then, was the drug not offered as a possible option (much less the best solution) for those at high risk for breast and ovarian cancer?

The trouble began in 1998, when American BCPT investigators stopped their tamoxifen trial. British scientists reacted quickly, and for the most part negatively, to the BCPT results. Although American scientists had stopped the trial for ethical reasons, because there seemed to be clear evidence of the drug's benefit, many of their British counterparts argued that the premature end of the trial was ethically problematic because it would be detrimental to all women in the long run. The president of the British Association of Surgical Oncology said: "I am shocked really. . . . At best, this is incompetence, and at worst, they are trying to get themselves publicity."[40] In particular, British scientists argued that ending the trial 14 months early would not help clinicians determine whether the drug prevented disease incidence or simply delayed it, or if there were significant long-term side effects that should be considered. The co-chairman of Britain's study of tamoxifen in high-risk women responded to the American trial as follows: "I think there has been a significant over-reaction from the US. . . . We desperately need to see what the long-term benefits of the follow-up will be. We need to see that these cancers are not coming back and that patients will not be dying of breast cancer."[41] These British scientists were reluctant to make a drug designed for long-term use in healthy women widely available on the basis of one study that had been prematurely stopped.

These British critics were particularly frustrated because they were engaged in a similar trial of their own and the Americans had, they felt, broken a promise to publish results of all of the studies together in 2000. The International Breast Cancer Intervention Study (IBIS), begun in 1994, which was coordinated in Britain but which involved women in Australia and elsewhere, was also testing the efficacy of tamoxifen among high-risk women. Trial organizers worried that, because the Americans had stopped the trial early, women would be less likely to enroll in their double-blind, placebo-controlled trial and more likely to visit a doctor and demand the drug. A surgeon who had been involved in organizing the trial predicted: "It is all going to take much longer now and it will be much more difficult to obtain worthwhile data."[42] One oncologist reported that patients began to request special appointments to ask for tamoxifen immediately after the American results were announced.[43] British researchers and physicians also argued that the drastic decision to prescribe a drug for long-term use in healthy women should be made only after careful consideration and with data from several studies. One of the IBIS organizers warned: "It behooves us all to get the right answer from these trials because there are a lot of women at stake."[44] British scientists and health-care professionals were reluctant to prescribe tamoxifen for healthy women on the basis of the results of one trial, but successfully completing other trials might prove very difficult.

The situation became even more complicated in July 1998, when researchers from the British IBIS trial and a similar Italian trial published early results that disputed the American findings. In articles published in *The Lancet*, their studies showed that tamoxifen did not reduce breast cancer incidence among high-risk women, and actually increased their risk of endometrial cancer. One article stated: "We have been unable to show any effect of tamoxifen on breast-cancer incidence in healthy women, contrary to the report from the NSABP-P1 study showing a 45 percent reduction in healthy women given tamoxifen versus placebo."[45] IBIS investigators suggested that the radically different results of IBIS and BCPT (the NSABP-P1 study) had to do with differences in the women who participated in the trial: "Differences in the study populations for the two trials may underlie these conflicting findings: eligibility in our trial was based predominantly on a strong family history of breast cancer whereas in the NSABP trial was mostly based on non-genetic risk factors. The importance of estrogen promotion may vary between such populations."[46] Indeed, IBIS had been even more restrictive than BCPT in restricting eligibility for their trial to subjects with extensive family histories of breast cancer. IBIS investigators concluded from these findings that there was not yet clear evidence of a preventive benefit of tamoxifen, and thus it should not be made available even to women at a very high risk for breast cancer. They continued to believe, however, that tamoxifen had some sort of protective effect, and thus required much more research, and particularly long-term study.

The findings of the IBIS trial did not get much press coverage in the United States and American NSABP investigators largely dismissed the results. The leader of BCPT argued that the British trial could not reach firm conclusions because it had a smaller sample size and therefore less statistical power, concluding that it didn't "shake any of our confidence in the findings of the US study."[47] Others argued that because the British focused on the effects of tamoxifen among women deemed to be at high risk because of family history, it was less useful in determining the drug's impact among women who were at higher risk because of a variety of factors (e.g., early age at first period, use of birth-control pills).[48]

Of course, IBIS's choice of a study population made it more useful for proponents of the British BRCA-testing system who saw extensive family history of breast cancer as an indicator of BRCA risk, and they chose not to prescribe tamoxifen for high-risk clients or for those who tested positive for a BRCA mutation. Even before the Americans stopped their tamoxifen trial, the consensus report written by PHGU had expressed caution about the possibility of using the drug as a chemopreventive: "A variety of agents, such as the anti-estrogen tamoxifen, have been suggested as candidate drugs which might reduce the risk of breast cancer. However, their use remains experimental, and the results of clinical trials to assess their effectiveness are awaited."[49] Their opinions changed little after the end of the US trial. In a newsletter dated December 2000, PHGU stated:

> It has been difficult, in the years since the discovery of the BRCA1 and BRCA2 genes, to know what treatment and/or prophylaxis to recommend for women carrying mutations in these genes as it has not been clear whether their cancers respond in the same way as the 95% of breast cancer cases not associated with these genes. Information is, however, now beginning to accumulate and the study of Narod et al. [regarding the effectiveness of tamoxifen in high-risk women] is a useful addition to the evidence base. It is interesting, but not yet explicable, that the protective effect of tamoxifen in their study was significantly greater in North American centers than in Europe.[50]

The British argued that tamoxifen had unclear utility in preventing breast cancer, particularly among clients with a family history of the disease, and should therefore be studied further before becoming part of the clinical recommendations for BRCA mutation-positive individuals. After the American trial was stopped, a British oncologist involved in BRCA testing responded: "The tamoxifen stuff I just dismiss because it's totally useless, unhelpful information. We're still running the study here, so one day we'll actually know if tamoxifen reduces mortality."[51] To him, the British still didn't know the effects of tamoxifen. British physicians were particularly apprehensive about prescribing a drug for long-term use that they felt produced equivocal benefits while increasing other risks. This approach to tamoxifen was only validated in the 2004 NICE guidelines, which simply noted that the drug hadn't been approved for use in the United Kingdom.[52]

As most British clinicians and scientists dismissed the American results, and as the government refused to license tamoxifen as a chemopreventive, there was little effort to convince them otherwise. While a Zeneca representative argued that the US study had "enormous statistical power and there was less than a one in 100,000 chance that it had produced freak results," Zeneca did not seek to shape the scientific controversy or to get drug approval for chemopreventive use from British and European regulatory agencies as it had in the United States.[53] Why not? Wouldn't the Britain-based company be as concerned with marketing its drug in its home country, a nation that reported a similar incidence of breast cancer, as it would be in the United States? In fact, the company chose not to pursue drug approval in Europe because of its patent position there— although the company had 4 years of patent protection left to cover tamoxifen in the United States, it had already run out of its patent protection covering the drug in Britain and the rest of Europe. Zeneca representatives argued that it would not make "commercial sense" to promote the drug in the region because the company would make little money while generic drug manufacturers took advantage of its marketing efforts.[54]

Why did British scientists, clinicians, government officials, and providers of BRCA tests take such a different approach to the chemopreventive use of tamoxifen than the Americans? It is possible that the British did not feel the same level of pressure to approve tamoxifen for widespread use as was felt by Myriad, clinicians, and the FDA in the United States. Tamoxifen was a chemopreventive that could help a newly defined group of at-risk clients in the United States, while the high-risk category, which was based on family-history information, was not a new one in Britain—it had been long established by breast cancer risk-assessment services in Britain and was already linked to a menu of management options. The drug might have also seemed more viable in the United States where the costs for the expensive prophylactic would be borne privately by clients and insurance companies, not the National Health Service. In Britain, NHS officials had to make population-wide calculations about whether the benefits of widespread use outweighed the financial costs and medical risks of the drug. The different approaches to the tamoxifen trial results might also be explained by the way providers' oriented their testing systems and understood the relationship between family history and BRCA-mutation status. The British sought to develop management options for clients who had a strong family history of breast or ovarian cancer and were thus considered high-risk and BRCA-mutation-positive. As a result, the equivocal results of the British tamoxifen trial were particularly valuable. Myriad, by contrast, problematized the link between family history and BRCA mutations and thus were less likely to endorse the British results. They were, therefore, more likely to agree with the positive American results that had a broader definition of "high-risk," which was not limited to family history.

Research into the effects of tamoxifen on women with BRCA mutations is ongoing. In a study published in 2001, Mary-Claire King and her colleagues found that tamoxifen reduced breast cancer incidence among healthy BRCA2 carriers, but did not affect those with BRCA1 mutations.[55] Researchers attributed this finding to the differences in estrogen receptivity of the tumors that these mutations help to produce. This was, however, a very small study that only involved 19 individuals, and did not seem to affect attitudes toward tamoxifen in either the United States or Britain.

Managing Moderate-Risk Clients

Builders of the British system sought not only to manage those deemed to be at high risk for breast and ovarian cancer, but also offered clinical interventions to those classified as moderate risk. Although these clients did not have a strong enough family history of breast and/or ovarian cancer to justify classification as high risk or to justify giving them access to BRCA testing, proponents argued that their family's history of breast and/or ovarian cancer was significant enough to warrant additional attention within a system devoted to ensuring the health of the entire public. At a meeting that focused on developing management strategies for clients with breast cancer risk, the University of Cambridge oncologist James Mackay argued: "Clearly genetic testing is only going to be of clinical importance to a small number of people. There must be a management strategy for those at significantly increased risk, in whom genetic testing is not feasible."[56] Health-care professionals affiliated with PHGU felt that they needed to develop some type of management approach to deal with these clients who had increased breast cancer risk but in whom a BRCA mutation was unlikely. Definition of this group of moderate-risk clients further demonstrated the British system's emphasis on the public's health—they wanted to identify as many people who were at elevated (even moderately elevated) risk of breast and ovarian cancer as could be identified.

James Mackay proposed that moderate-risk clients be offered access to an age-related mammographic screening study that would investigate clinical effectiveness of increased management of such women. Others agreed with this strategy. "In contrast to high risk women," one oncologist concluded, "and given the lack of evidence of benefit for any intervention for

Table 4.1

Management of women at moderate risk. Source: James Mackay et al., "Familial Breast Cancer: Managing the Risk" (Anglia Clinical Audit and Effectiveness Team, 1997).

Age	Management options available
Below 30	No mammography
30–34 Youngest affected first-degree relative diagnosed age 40+	No mammography
Youngest affected first-degree relative diagnosed age 39 or below	Annual mammographic screening starts 5 years below the age of diagnosis of the youngest affected relative
35–49	Annual mammography and annual clinical examination
50 and over	Mammography every 18 months with clinical examination around time of mammography; half of these mammograms will be performed within the National Breast Screening Programme

women in this group . . . an intervention should only be offered to moderate risk women in the context of a national research study."[57]

Using published data on the effectiveness of mammography in reducing breast cancer incidence and specifically, "mortality reduction," Mackay and his colleagues constructed the scheme illustrated here in table 4.1. Researchers were, however, originally quite concerned about the ethics of a randomized controlled study.[58] Indeed, one could imagine that proposing to study the effectiveness of the mammographic technology that had already become an accepted part of women's health care through a randomized controlled study would be particularly controversial; in the United States, for example, recent studies questioning the effectiveness of mammography in decreasing mortality have been received with skepticism on the one hand and numerous testimonials about the technology's effectiveness in detecting breast cancer at its very early stages on the other.[59] Mackay's study, however, did receive funding from a private UK medical charity, and half of the women defined as moderate-risk were offered age-related mammography screening in order to determine the most effective treatments available for women not eligible for genetic testing. Classifying a client as moderate-risk kept her from getting access to mastectomy, but it granted consideration for the mammographic screening study. Thus, the classification of a client as moderate-risk allowed her increased access to health-care options in comparison to a low-risk client.

This approach was validated in the 2004 NICE guidelines, which suggested that moderate-risk clients aged 40–49 years have access to annual mammographic surveillance, while moderate-risk clients aged 30–39 years could enroll in research protocols to investigate the efficacy of mammographic screening in women of this age. (Women of age 50 and older already had access to annual mammography.) All were to be cared for at the secondary-care level (e.g., by oncologists or breast surgeons) rather than at the genetics clinic. This attention to clients in the moderate-risk category, in particular, illuminates again the purpose and orientation of the British BRCA-testing system. Indeed, it wasn't a BRCA-testing system at all; In Britain, in contrast to the United States, the testing system reflected the NHS's public health goals and was devoted to identifying all those who had a familial risk of breast or ovarian cancer, regardless of BRCA mutation, and providing them with management options that would be beneficial for their overall health.

Conclusion

The BRCA-testing systems that were built and integrated into health care in the United States and in Britain defined risk, disease, and the treatment options available in rather different ways. The creation of an at-risk individual marked by her DNA status seems, contrary to the predictions of commentators, not to be an automatic product of genetic medicine but rather the consequence of a specific technological architecture. Although in the United States the critics of geneticization appear to be correct, as Myriad created new risk categories and defined diseases and treatments according to gene mutation status, in Britain risk categories were defined by family history and NHS officials were reluctant to create novel diseases or authorize risky new treatments simply because of the availability of genetic testing. These differences reflected distinct definitions of a good health outcome. In the United States, Myriad sought to identify and find treatments for as many BRCA mutation-positive clients as was possible. In contrast, the British national system used the system to ensure good health throughout the population, by identifying all those deemed to be at risk for breast and ovarian cancer and suggesting treatments that would maximize prevention.

There were also significant differences in how access to management options was controlled. Those differences mirrored the different ways in which access to BRCA testing had been shaped in the two countries. In the United States, while Myriad suggested the availability of tamoxifen, prophylactic surgery, and increased surveillance for BRCA mutation-positive clients, it did not have—nor did it want—any control over what medical interventions were actually taken. Decisions about which option (if any) to choose were made entirely by the physician and the client. In Britain, proponents of the national standard (and, later, NICE) had considerably more control. Adherence to the NICE guidelines was necessary for continued funding, and thus regional health-care professionals had only limited authority in directing the care of their clients. Moreover, tamoxifen had not been approved for use as a chemopreventive agent in Britain. Clients using the British system, then, could not demand access to BRCA testing or tamoxifen, and health-care professionals could not choose to offer them either. While these decisions had been left up to the health-care professional and the client in the United States, their counterparts in Britain did not have similar control over the medical interventions used.

We can also see, as was discussed in the last three chapters, that there was a clear difference in the way the technologies in the two countries approached the issue of client demand, both in terms of the architectures of the systems and the management options that were made available. In the United States, Myriad encouraged demand for BRCA testing; In fact, it argued that more people should be tested because they would receive better health care as a result. In Britain, patient demand was seen as a problem. The more people who called their doctors demanding access to testing or chemoprevention, the less time and money physicians would have to treat those who really needed care. British physicians expressed such sentiments when the Americans stopped their tamoxifen trial. One noted: "The Americans have unblinded the trial, which means it will be unbalanced and they will not be able to answer many questions. . . . You can't treat everybody in the world with tamoxifen. You must identify who is going to benefit and who is not."[60] To these British physicians, it was the doctor's responsibility, not the patient's, to identify need and the appropriate course of medical action.

To what extent did these technologies and their differing approaches to risk, disease, and treatment influence the discussions that took place between doctor and patient? It would be safe to assume that in the United States Myriad's test results, which classified clients into risk categories, and Myriad's promotional and educational materials, probably shaped the discussion between health-care professionals and clients in terms of their risk and disease status as well as the treatment options available. It is important to remember, however, that Myriad provided health-care professionals with considerable autonomy, and thus they could counsel clients however they wished. They could, for example, inform clients about the negative side effects of tamoxifen or strongly encourage clients to have a family member affected by breast or ovarian cancer tested first so that they could gain a better understanding of the meaning of a gene mutation in terms of disease incidence. Of course, as mentioned in previous chapters, clients influenced by Myriad's marketing materials could always seek their desired medical intervention elsewhere, through another physician. Furthermore, studies suggest that physicians are often not well informed about breast cancer genetics or BRCA testing, which would surely impair their ability to counsel clients both before and after DNA analysis.[61]

In Britain, approaches to risk, disease, and treatment were codified in the NICE guidelines and drove the funding decisions of the NHS administration. Thus, the national system was likely to define and constrain the behavior of health-care professionals at the primary and secondary-care level, as well as at regional genetics clinics. Clients who sought to circumvent these approaches to BRCA testing or to gain access to additional risk management options would have to seek care far outside the country. As of 2006, most European countries had developed risk assessment and triage schemes similar to the one in Britain and had not approved tamoxifen. A very small number of British clients circumvented the national NHS system by using Myriad's BRCA-testing system, but of course they could not get access to tamoxifen as a chemopreventive through any European physician (except through ongoing clinical trials).[62]

At this stage in the comparison, it seems worthwhile to begin to ask "Which system is better?" This chapter has shown that the answer to this question is, at the very least, extremely complicated. Determining which system is better requires us to first develop a way to measure success or failure. But whose measure do we use? Do we adopt Myriad's goal of identifying and treating as many BRCA mutation-positive people as is possible? Or do we judge the system according to the National Health Service's goals of identifying and managing everyone with a family history of breast or ovarian cancer? Should the system be oriented toward offering greater "consumer choice" (as in the United States), or toward offering greater "common good" (as in Britain)? BRCA testing in each country reflected distinct understandings of good health care and had specific and different implications for those who used the new technology. It is difficult, if not impossible, to judge the benefits of a technology when we cannot agree on what functions are most beneficial and when it is clear that our assessments of benefit are tied to our national approaches to health care, commercialization, and individual empowerment. As more and more technologies are transported across national borders, however, such questions are being increasingly raised. In the next chapter, we will see whether and how they were resolved in the case of Myriad's attempt to transfer its BRACAnalysis to Britain.

5 Myriad, Britain, and Culture Clash

Once systems of genetic testing for breast cancer had been built and integrated into health care in the United States and Britain, Myriad Genetics attempted to expand its testing service and its sources of revenue to Canada, Europe, South America, and Asia.[1] The first target of these efforts was Britain. Perhaps Myriad assumed that it would be simple to transfer its technology to a country that had reported incidences of breast and ovarian cancer similar to those in the United States and seemed to have a strong commitment to genetic medicine. Hoping to shut down the National Health Service's BRCA-testing system and to have the blood samples that were collected in Britain analyzed at its laboratory in Salt Lake City, Myriad began trying to convince health-care professionals of the value of its technology and threatening legal action (on the grounds of patent infringement) against those who continued to provide testing.

What happened when Myriad tried to expand its version of BRCA-testing services to Britain and the rest of Europe? Would Myriad's technology and way of structuring the identities of clients and health-care professionals be neatly transferable to the British context? And how would the British government, British scientists, British clinicians, and British patients respond to Myriad's attempt to insinuate its approach to BRCA testing into their institutions? Would they be able to accept Myriad's definition of a good health outcome, or would they remain loyal to the British approach?

This chapter explores Myriad's attempt to develop a transnational service of genetic testing for breast cancer, and how British scientists, health-care professionals, and activists responded to these efforts. As will be evident, ongoing tensions emerged, for in working to extend its patent rights Myriad was not simply trying to introduce a single entity of narrow scope into a new geographic region; it was trying to introduce an entire system—encompassing the clinical and technical dimensions of the test, particular roles for testing system participants such as clients and health-care professionals, and specific definitions of a good health outcome—into a cultural context that differed greatly from the one in which it had been built.

The chapter begins by describing how Myriad attempted to expand its testing service to Britain. It then explores how British scientists, health-care professionals, and activists challenged Myriad in three ways: they questioned its use of patent rights as a justification for expansion of its testing service; they disputed the validity of a BRCA-testing system focused on laboratory services; and they argued that Myriad's system prescribed roles for health-care professionals and clients that were inappropriate in the British context. Finally, the chapter describes the negotiations between Myriad and the National Health Service and the eventual resolution of Myriad's attempt at technology transfer.

Myriad Tries to Transfer Its Technology

When Myriad turned its attention to the international market, it adopted an approach similar to its expansion efforts in the United States—using its legal and economic position to eliminate its competitors. Indeed, Myriad had already applied for patents covering the BRCA genes and resulting diagnostics and therapeutics from the European Patent Office, which covers most countries in Europe, when it applied for American patents in 1994 and 1995.[2] By 1998, expecting that the European Patent Office would soon grant its patent applications, it began to market its BRCA-testing regime directly to European health-care professionals. Its strategy was to emphasize that it could provide an accurate laboratory service that would be widely available.

Myriad began its European expansion efforts in earnest by inviting representatives from the European Familial Breast Cancer Demonstration Project (an initiative designed to investigate methods of management for women at high risk for breast cancer) to tour its laboratories and facilities in 1998. Company officials argued that cooperation between Myriad and the European Project would allow for an expansion of the "currently limited availability of breast cancer genetic testing in Europe," emphasizing both the superiority of their DNA sequencing and the availability of their services to a much larger clientele than those eligible for BRCA testing under most European state-run health systems.[3] It hoped to convince Project members, which included delegates from the United Kingdom, France, Italy, Germany, Norway, and the Netherlands, that it could provide better services that were more technically sophisticated than those already available in those countries. As was discussed in chapter 1, Myriad's BRCA-testing services involved full sequencing of both BRCA genes, which was considered more analytically sensitive than most European services that used a variety of DNA analysis methods.[4] At the time, Dr. Neva Haites, head of the European project, approached the visit in a positive manner: "This meeting offers us the possibility of understanding the potential for future collaborations between European centers and Myriad Genetics Laboratories, to ensure that high-risk clients can be identified and hence are offered optimal screening and management."[5] Despite Haites's initial enthusiasm, however, few European health-care professionals seemed interested in using Myriad's services. Most seemed to prefer to continue with their existing national systems of BRCA testing.

By the end of 1998, Myriad had focused its efforts on Britain and settled on a strategy that had been successful both for it and many others engaged in medical genetics and biotechnology in the United States: threatening legal action on grounds of patent infringement.[6] The company's chief executive officer and its lawyers presented their case to a biennial meeting of the UK Cancer Family Study Group, likely unaware that its members had played a pivotal early role in developing the national BRCA-testing system in Britain. As it had in the United States, Myriad argued that the British NHS, by providing BRCA testing, would be in violation of Myriad's European patents as soon as they were issued, just as the University of Pennsylvania, the Genetics and IVF Institute, and Oncormed had violated its US patents. British regional genetics clinics, company representatives demanded, must begin sending their samples to its US laboratories or else face a lawsuit. British health-care professionals, however, were not moved by Myriad's promise of a better testing service or by its threats to file suit and shut down NHS BRCA-testing services. They continued to provide testing as a national system that included risk assessment, triage, and an integrated package of counseling and laboratory analysis services.

Myriad then tried another approach: it directly contacted the UK Department of Health, which was in charge of NHS services. It demanded that the UK DoH either pay a licensing fee to continue testing, send samples to Myriad's US laboratories, or risk getting sued for patent infringement.[7] Meanwhile, Myriad also explored other options—for example, it contacted private laboratories in Britain to see if they were willing to serve as satellite laboratories to process British samples and then send mutation information and a share of the revenues back to Myriad's BRCA-gene database in Salt Lake City.

Responding to Myriad

Myriad's concerted effort to pressure the UK Department of Health and health-care professionals to adopt its testing service led British scientists, health-care professionals, activists, clinicians, and government officials to begin organizing targeted responses. The Clinical Molecular Genetics Society (which had been involved in developing guidelines for BRCA-testing services in Britain) and the British Society of Human Genetics began to compose position papers and official statements questioning the patentability of genes and predicting negative consequences for research and health care if human gene sequences were allowed to be owned. The patient activist Wendy Watson also gave interviews to the media expressing her concern over gene patenting and the commercialization of genetic testing and helped to mobilize opposition to Myriad among patient groups.[8] Meanwhile, the UK DoH developed a consultation committee to help it in its discussions with Myriad, which included, along with Wendy Watson, the chairperson of the Clinical Molecular Genetics Society; physicians, counselors, and nurses from regional genetics clinics; and NHS officials involved in purchasing regional services.[9]

The Legitimacy of Patent Rights

In the United States, acquisition of patent rights over the BRCA1 and BRCA2 genes by Myriad had not only identified the company as an inventor of the isolated and purified genes but also helped justify its efforts to become the sole provider of BRCA testing and control how testing would be built. While scientists, health-care professionals, and activists in the United States had questioned the architecture of Myriad's testing system and the roles it prescribed for health-care professionals and clients, they did not initially challenge assignment of inventorship of the BRCA genes to Myriad, or the ownership rights that these patents provided.[10] In fact, there had been little organized opposition to the patenting of disease genes at all. Although the American Society of Human Genetics had intervened in Craig Venter's attempt to patent DNA sequences of unknown utility during the Human Genome Project, it was quiet on the patenting of disease genes.[11] Scientists were likely reluctant to speak because they were accustomed to a research culture that encouraged them to capitalize on their work either by patenting inventions with the help of universities' technology-transfer offices or starting small companies of their own.[12] Scholars of bioethics and the law initially had a subdued response to the prospects of patenting disease genes as well, although they began to worry about the practice soon after the BRCA genes were patented. Meanwhile, patient advocates seemed ambivalent on the issue, worrying about the ownership of life but also accepting the argument that gene patents were necessary to stimulate innovation.[13] National discussion and activism about the issue did eventually emerge by the early 2000s, but it took quite a long time and was on a relatively modest scale, particularly in comparison to Europe.[14]

This limited opposition within the scientific, patient, and bioethics communities might be better understood by considering the legal, regulatory and industrial environments in the United States. As was discussed in chapter 1, not only were linkages between the university and industrial sectors becoming more common; technology-transfer offices at American universities actively encouraged their scientists to patent their inventions and some scientists even left academia to start companies and commercialize their own research findings.[15] Furthermore, American courts had a history of privileging the property and economic interests of biotechnology researchers. In the landmark 1980 case *Diamond v. Chakrabarty*, the US Supreme Court held that living organisms were patentable. In 1990, the Supreme Court of California decided, in *Moore v. Regents of the University of California*, that a cancer patient named John Moore did not own his cells, in part because giving him such a right would "hinder research by restricting access to the necessary raw materials."[16] The idea that researchers, even those in the academy, could own and commercialize biotechnological inventions was well accepted by 1997–98, when the BRCA patents were granted in the United States.

Would patents covering gene sequences have the same meaning in Britain? A number of commentators have suggested that understandings and attributions of inventorship and ownership of scientific and technological objects are contingent upon social context. Stephen Hilgartner describes how laboratories participating in the project to map and sequence the human genome defined property rights and the distinctions between public and private domains in very specific ways.[17] Locally contingent definitions of inventorship and property are also evident in how patent rights are attributed and used. Marianne de Laet argues that patents "are different things in different places," noting that a single patent can be, simultaneously, a recognition of achievement at a laboratory in the Netherlands, a mechanism to protect innovation at the World Intellectual Property Organization in Switzerland, and a source of information at a government ministry in Africa.[18]

Many aspects of British and European patent law, university-industry relations, and health care suggested that the BRCA patents and the resulting monopolies on BRCA testing would be understood quite differently in Europe than in the United States. In contrast to their American counterparts, European universities did not so actively encourage their scientists to patent their work and had not had such a historically close relationship with the industrial sector. Very few European scientists left academia to start companies.[19] In addition, although American, British, and European patent laws look similar in most respects, there are three important differences that are directly relevant to the patentability of disease genes. First, the European Patent Convention (which governs the European Patent Office) and the UK patent laws (which govern the UK Patent Office), unlike their American counterparts, explicitly prohibit patent protections for methods of medical treatment as well as inventions that are considered contrary to *ordre public* (public order or morality). Second, the European Patent Office has an opposition mechanism that allows anyone, from an advocacy group to a government, to oppose a patent, on the grounds of its issue (capable of industrial application, novelty, and inventive step) or that it is exempted from patentability (e.g., it was an affront to *ordre public*). The mechanism and potential grounds for opposition suggest that public health and policy concerns are important to determinations of patentability, whereas the granting of patents in the United States are seen primarily as a technical enterprise of interest to inventors (usually scientists and engineers), their lawyers, and the US Patent and Trademark Office. Third, there was already some evidence that biotechnology patents would be extremely controversial in Europe. A patent on the Harvard Oncomouse, a mouse that had been genetically engineered to carry a specific gene that would cause cancer (called an oncogene), had been the subject of much public debate throughout Europe, as many argued that it was morally wrong to patent or own life itself. In fact, the European Patent Office initially rejected the patent application in 1989 because examiners felt that it violated the European Patent Convention's prohibition on patenting animals. Fifteen years after the initial rejection, and after considerable discussion in the media and by governments and opposition proceedings initiated by environmental and animal-welfare groups at the European Patent Office, a restricted patent on the invention was granted.[20] Although British and European patent laws and the Oncomouse controversy did not figure directly in Myriad's proposed technology transfer to Britain, the concerns and priorities they raised set the stage for the controversy that ensued.

In fact, British scientists, health-care professionals, and activists had begun to mobilize against the patenting of genes long before Myriad tried to expand its testing service, in response to European Union legislation designed to strengthen the European biotechnology industry by harmonizing patent law across member countries.[21] The EU Directive on the Legal Protection of Biotechnological Invention (hereafter referred to as the Biotech Patent Directive), first introduced in the European Parliament in 1988, aimed to make human gene sequences, as well as genetically engineered plants and animals, patentable across the European Union. A common, robust patent law was seen as pivotal for the European biotechnology industry, which many hoped would contribute to the growth of the European common market. Because of the concurrent debate about the patentability of the Oncomouse, the European Parliament became increasingly interested in reviewing the ethical dimensions of the proposed law. Parliamentarians dealt with these ethical concerns by proposing a number of amendments (one of which clarified the public-morality exemption as it related to biotechnology patents), but in 1995 the law was rejected. It was almost immediately reintroduced in the Parliament, but by then it had become extremely controversial throughout Europe, inspiring activism from governments, environmental, religious, and development nongovernmental organizations, and patient, scientific, and medical communities, as well as the pharmaceutical and biotechnology industries. Here too, patentability was seen as a social and political, rather than purely technical, issue.

Opponents of the directive, which included voices as diverse as Greenpeace, Danish Patients with Genetic Disorders, and feminist groups, criticized the Biotech Patent Directive for a variety of reasons. They questioned the ownership and commercialization of what they considered to be fundamental building blocks of life, which some felt was inherently immoral and others argued would have negative implications for the health, environment, and economies of both developing and developed countries. Some criticized the patenting of human gene sequences in particular, arguing that they were not patentable because they were discoveries of things already existing in the body rather than inventions of novel things that could be subjected to rules of intellectual property.[22]

A number of British groups were actively involved in this opposition to the Biotech Patent Directive. For example, in September 1997 the British Society for Human Genetics (BSHG) issued a statement, titled "Patenting of Human Gene Sequences and the EU Draft Directive," which argued that the EU Directive should not be passed, because genes of both known (such as those linked to disease) and unknown utility did not fulfill the first criteria of patenting, novelty, and were therefore unpatentable: "[Novelty] cannot reside in the mere description of a nucleotide sequence. It must rest in either novel methodologies for discovering the sequence or a novel use or application of the sequence."[23] Simply identifying an existing nucleotide sequence, the BSHG argued, did not require ingenuity on the part of the researcher. A number of British geneticists also signed a separate letter to the EU Parliament that articulated these same concerns. The letter began: "As researchers or clinical scientists we urge you to exempt genes and their elements from patenting. . . ."[24] It was the first time that British geneticists had made this type of concerted grassroots effort to influence policymaking, and they had come together to question the attribution of ownership that had certified Myriad's legitimacy in the United States.

The BSHG and similar organizations were not uniformly against the practice of patenting. In fact, the BSHG's 1997 statement on the subject called patenting "a valuable means of protecting intellectual property and promoting investment in developing products for the diagnosis and treatment of genetic disease."[25] These researchers suggested that although patenting genetic diagnostics and therapeutics (including specific systems for genetic testing) was entirely understandable, patenting genes themselves would be detrimental to the cultures of research and health care in Britain and the rest of Europe. They worried that assigning ownership for gene sequences would allow patent holders to control all research on a particular gene for the life of the patent, potentially limiting research opportunities and preventing scientists from working on the most lucrative and complex biomedical problems. They also wondered if a focus on intellectual property and commercialization of medical biotechnology would conflict with European conceptions of health care as a public good.

Wendy Watson, who also campaigned against the EU Directive, had a perspective similar to that of the BSHG. "You can't patent a gene," she asserted, "a gene is a discovery! It's not an invention, it's a discovery!"[26] Even the Church of Scotland issued a statement in opposition of the Biotech Patent Directive, declaring that "living organisms and genetic material of human origin are in themselves unpatentable, as parts of God's creation."[27] These groups questioned the very idea that genes were patentable inventions and by extension, the basis of Myriad's effort to expand its testing service to the United Kingdom and shut down the National Health Service's BRCA-testing service.

These critics did not succeed in stopping the EU legislation. In May 1998, after years of vigorous debate, the EU Parliament and Commission passed the Biotech Patent Directive, which allowed the patenting of, among other things, genes that had been isolated from the body.[28] Despite this law, which seemed to allow patenting of human genes, controversy continued over whether human genes should be patentable and what kind of ramifications patenting would create for European health-care systems. In fact, a number of countries immediately challenged the Biotech Patent Directive in the European courts.[29]

Questioning the Patentability of the BRCA Genes

The controversy over the Biotech Patent Directive at the EU level had important consequences for the British debate regarding Myriad's expansion campaign. The directive seemed to justify Myriad's patents, but the new law had also sparked tremendous resistance among many groups in Britain and galvanized critics to articulate a clear position against the patenting of genes—Myriad's major claim for the legitimacy of its technology transfer.

It should come as no surprise, then, that British opposition to gene patenting continued and even grew after the Biotech Patent Directive was passed, or that it soon focused on Myriad. The lines of argument echoed those that had been used against the Biotech Patent Directive. First, opponents argued that no disease genes should be patented, both because they were part of nature and because ownership of gene sequences would be detrimental to both research and health care. Second, they argued that attributions of ownership were particularly complicated in genetics research, and that serious discussion was needed on the definition of the "inventor" before making genes patentable.

Just as Myriad had begun its attempts in Britain, and just as the final negotiations on the Biotech Patent Directive were beginning, the BSHG issued a statement that was clearly directed toward Myriad. It asserted that "a natural human gene sequence is part of the human body, and as such should not be patentable." Furthermore, "the suggestion that such a sequence might be patentable if it is 'isolated in a pure form' or 'isolated outside of the body' seems to us a sophistry, and should not be allowed."[30] Although it had already published a statement opposing the legalization of gene patenting through the Biotech Patent Directive, the BSHG deemed the issue, and Myriad's attempted technology transfer, important enough to articulate its position again so that it might have a greater influence on the national debate.

The BSHG and other organizations used Myriad's attempt at European expansion to exemplify the dangers of patenting. The BSHG argued the holder of a patent on a gene would have the power to control all future research on that gene, and thereby, perhaps, to stifle innovation: "There is only one consensus of normal human sequence. If the sequence as such is patentable, it will not be possible for anyone at any time to devise a better or different way of genetic diagnosis; this is inequitable."[31] In a 1999 paper titled "Gene Patents and Clinical Molecular Genetic Testing in the UK," the Clinical Molecular Genetics Society (CMGS) echoed the BSHG's concerns. That paper predicted that allowing the holders of patents on genes to determine the provision of genetic testing would make genetics services prohibitively expensive, would reduce the availability of testing through the NHS, and would jeopardize clinical and laboratory expertise in the NHS by allowing private concerns to provide state-of-the-art services: "This development will be followed by other attempts to enforce gene patent rights. As such it raises issues on the most effective way of delivering patient care, forms a crossroads for molecular genetic testing and an important point for the development of government policy."[32] To these groups, questions of patentability could not be separated from questions of equity and of the quality of patient care.

Further, British scientists, health-care professionals, and activists asserted that, even if genes were inventions that could be patented and owned, discovering the BRCA genes entailed a collective effort involving researchers, women, and funding bodies in Britain as well as in the United States. Myriad simply could not claim to be the sole "inventor" of the BRCA genes.[33] Many British researchers argued that, if authorship of the genes could be claimed at all, they deserved some ownership because they too had contributed to the gene discoveries. Sir Walter Bodmer, a scientist who had been involved in early research on the BRCA1 gene, said: "Myriad is claiming it contributed far more than it actually achieved. As a result . . . there is a lot of feeling of unfairness among British scientists."[34] Other scientists simply argued that the BRCA gene discoveries were the result of a protracted collective effort, and the final mapping and sequencing was more a matter of luck than inventiveness. Andrew Read, chairman of the British Society of Human Genetics, explained: "The whole area of gene patenting is controversial because it gives the prize to the person who put the last brick in the wall. . . ."[35] British scientists frequently used this type of metaphor to explain their opposition to gene patenting, tapping into an age-old image of science as both disinterested and collective.[36] "We are uneasy," the oncologist Bruce Ponder explained, "about the principle of patenting genes. Finding a gene is just the final step in a pyramid of knowledge and the question is whether it is justifiable for one company to own the patent. . . ."[37] Many of these geneticists argued that because the gene discoveries were the result of considerable research done by multiple investigators across the world, the attribution of sole inventorship to Myriad simply did not make sense. Patient activists agreed. "I do know," Wendy Watson concluded, "that when it got to this stage, it was pure spade work, there was nothing inventive about it, it was pure spade work."[38] This sense of outrage, of course, contrasted starkly with the initial silence of American geneticists and activists with regard to Myriad's claims to inventorship and rightful ownership of BRCA testing in the United States.

Some geneticists pointed out that Myriad's claims to sole ownership were particularly offensive because most Britons (as well as most other

Europeans and Americans) credited Mike Stratton, not Myriad, with finding the BRCA2 gene. Establishing priority in the BRCA gene discoveries had been very controversial. The public excitement and potential scientific, medical, and industrial rewards had led a number of scientists to search for the genes and many even referred to the research as a "race."[39] Although the "winner" of the race to find BRCA1 (Myriad) had been relatively clear, the winner of the race to find and sequence BRCA2 had been much more difficult to determine.[40] The day before Mike Stratton's group published the BRCA2 gene sequence in the magazine *Nature*, Myriad announced that it had found the gene and submitted its sequence to GenBank, an international depository of gene sequence information. Both Myriad and Mike Stratton's group filed for US and European patents on the BRCA2 gene within days of one another, each claiming that they had mapped and sequenced the gene first. None of these patents had been issued when Myriad attempted its transnational technology transfer. This BRCA2 controversy led many of the British scientists and health-care professionals in Britain's cancer genetics community to feel aggrieved by Myriad's proposed expansion. One scientist said that she would rather continue testing and go to jail for patent infringement than accept Myriad's patent claims over the BRCA genes: "At the end of the day, I hope I am locked up, because I'll make such a big deal about it. I mean they say they'll try and enforce this patent but I just hope the NHS doesn't just cave in and pay them money. The other thing is that my [friend] found BRCA2 at Sutton [the Institute for Cancer Research]. So you can imagine how galling that is."[41] This scientist saw her proposed protest as an important political opportunity to stand up for her values, even comparing herself to her grandmother who was jailed for being a suffragette.

Finally, some scientists argued not only that the BRCA gene discovery was a product of multiple inventors, but also that the process of finding the gene was identical to that by which all other genes had been found. From their perspective, there was not even anything novel about the process of finding the BRCA genes; the process of gene discovery was a well-understood, widely used, and fairly uniform process. Any scientist engaged in the process of looking for any gene would have followed a process similar to Myriad's. The BSHG noted simply: "The discovery of gene sequence has for some little time been a well understood process. There is nothing novel or inventive about this in principle, and as such new gene sequences should not be patentable, even where a straightforward utility e.g. diagnostic testing has been specified, unless there has been real progress towards the design of a specific commercial product."[42] One scientist who was involved in breast cancer genetic research in Britain remarked: "Most of us are pretty uncomfortable about [patenting]. That finding the BRCA1 gene, in our view, didn't involve anything really novel. It's novel in the sense that they didn't know it was the BRCA1 gene until it was found, but it was a totally predictable consequence of the work that everybody was doing and there wasn't any particular reason why Myriad should scoop that particular pool, whether they were going to make a lot of money or not. It just didn't seem, to us, that they fulfilled the criteria of originality and so forth that you need for a patent."[43] British scientists simply weren't prepared to accept that there was anything novel or unique in finding the BRCA genes that deserved the attribution of sole inventorship and ownership to Myriad.

Myriad executives responded to these criticisms in the British media, arguing that Myriad deserved the title of inventor and the accompanying benefit because of the time and money it had spent: "We've invested an enormous amount of man-years in making this discovery and making it applicable. It's only right that we should be protected." But while Myriad based its claim to ownership on the resources it had spent mapping and sequencing the BRCA genes, British critics also suggested a broader definition of resources that included donated blood from families with histories of breast and ovarian cancer, studies that contributed to the overall body of knowledge about breast cancer genetics, and money from groups that had funded research that had led to the discoveries. One scientist involved in the BRCA gene research noted that it would be unfortunate if those who helped to find the BRCA genes by donating blood samples would later have to pay Myriad to receive access to testing.[44] Wendy Watson echoed this sentiment: "Nobody has the right to patent this kind of information, which was only found with the help of the many families who had suffered a case of hereditary cancer. . . . It is morally wrong that any company should benefit commercially from that kind of research."[45] Wendy Watson also took this position further, arguing that it was not only Myriad's money that had contributed to finding the BRCA genes, but also money from UK medical charities: "It was charity money that was looking for the gene, academic money, not private enterprise money that was looking for the gene."[46] When charity money was involved in research, Watson suggested, either the resulting technologies should not be patented or else revenues gained through such ownership should be funneled into more research. This was quite a different environment from the United States, where the government had decided in the 1980 Bayh-Dole Act not to assert any property interest in the research it funded. These commentators argued that, whereas in the United States scientists and other testing providers simply accepted that Myriad's patent rights gave it control over the provision of testing, the BRCA genes were the result of multiple contributions from a variety of sources, and that Myriad had no right to claim sole ownership or control.

Opposing the Architecture of Myriad's Testing System

British critics also responded to Myriad's proposed expansion by challenging the appropriateness of Myriad's system in a context in which health care was provided by the state, and clinical care was traditionally integrated with laboratory services. They attacked the architecture of the system, arguing that Myriad's focus on DNA sequencing missed the importance of both counseling about the uncertainty of genetic risk and linking technical results to preventive care. They also disagreed with the roles of health-care professionals and clients prescribed by Myriad's system, arguing that clients should not be consumers whose health-care decisions were simply facilitated by clinicians.

In contrast to Myriad, the National Health Service had built a BRCA-testing system based on risk assessment and triage that integrated laboratory analysis and clinical care, demonstrated a commitment to genetic counseling, and was well integrated into existing practices of genetics and health care. One molecular geneticist suggested directly that the testing-counseling combination was part of the British ethos:

We have quite a strong ethic, in this country, that suggests that many types of genetic tests should be coupled to access to genetic counseling, in fact it should be a package. . . . pre-test counseling, the test, and then post-test counseling, and then maybe, long term follow-up after that for some individuals. Or at least the possibility that they can come back if they're still worried. So it's part of a continuing care package, the genetic test, the technical test is only in some ways, the easiest part of it. And we wouldn't like to see genetic testing decoupled from access to counseling. And in fact, it may be that genetic testing became discredited if it were decoupled from access to counseling so that's something that we worry about and are quite keen to preserve.[47]

If British physicians were forced to send their samples to Myriad's laboratories in the United States or to Myriad-approved laboratories in Britain, patients would no longer have to go through regional genetics services to gain access to laboratory analysis, and there would be no guarantee that they would receive appropriate counseling from NHS-trained health-care professionals. Myriad's system, critics suggested, would both destroy their testing system and jeopardize the NHS's commitment to providing a package of genetics services. The CMGS report articulated this concern:

Focusing genetic testing in multidisciplinary Regional Genetic Centres assures the link between diagnosis and counseling that is the hallmark and assurance of quality in this area of Medicine. . . . The 'testing-within-counseling' culture may be lost if the laboratories are divorced from the counsellors. At worst, a group of patients and families for whom a genetic diagnosis is made will be at risk from the consequences of weak counseling and may be lost to key follow-up systems.[48]

Myriad's opponents also argued that while the company claimed to have a comprehensive testing system that used the best available techniques to fully sequence both genes, it was not accurate for their purposes.[49] For them, as was discussed in chapter 4, finding a deleterious mutation that could facilitate managing the health care of women at risk for breast or ovarian cancer was much more important than producing genomic information by fully sequencing both BRCA genes. Most British geneticists did not apologize for this approach. First, if clients no longer had to use the NHS's system of risk assessment and triage to gain access to testing, health-care professionals would lose their opportunity to identify and monitor those at moderate risk. Second, British geneticists suggested that their system made the best use of limited resources. One said: "We're not as effective as Myriad, because we don't sequence the gene in quite the same way, but we do it in a more intellectual fashion. . . . If we came to a mutation, we'd stop, whereas they sequence the whole gene."[50] Such an approach also allowed British providers of genetic testing to cut costs and offer more tests and manage more clients within a limited budget: an important consideration in a government-run system focused on preventive care. In addition, British opponents criticized the technical accuracy of Myriad's methods of DNA analysis by noting that the "gold-standard" method of DNA sequencing that Myriad used would not necessarily detect all the alterations in the BRCA genes. "We do not question the validity of the idea that a sequencing test is currently a technical 'gold standard,'" said one molecular geneticist. "Sequencing is probably not 100 percent sensitive though and the Myriad test does not claim to pick up all mutations—deletions for instance—and cannot pick up mutations in other breast cancer predisposition genes apart from BRCA1 [and] 2."[51] If Myriad's method of laboratory analysis wasn't even 100 percent sensitive in picking up all mutations, critics argued, it certainly didn't warrant relinquishing an approach that enhanced identification of inherited risk through assessment of family history.

This dispute over the accuracy of BRCA testing is not particularly surprising. Commentators have long argued that the determination of accuracy for any test is achieved through a series of social and political decisions.[52] Developers must agree on the desired outcome, on methods for measuring performance, and even on the objects of focus. In the case of BRCA testing, the priorities of test developers—public health in Britain, availability of a state-of-the-art product for consumers in the United States—had led to different agreements about the definition of the genetic testing technology, the value of generating genomic information, and even the utility of finding BRCA mutations in preventing cancer, and thus the architecture of an accurate genetic test.

Not only did the components of Myriad's testing system come under fire; British health-care professionals also questioned whether the roles of the system participants articulated in the company's testing system were appropriate in the British context. They argued that Myriad's system would remove the gatekeeping authority of health-care professionals and possibly jeopardize the future of genetic medicine in Britain. The British BRCA-testing system provided genetics clinics with the authority to direct care while demonstrating to administrators that the NHS could provide genetics services for common diseases within the existing NHS culture. By creating an independent laboratory that would allow clients to circumvent the system of risk assessment and triage, Myriad's system would diminish the authority of health-care professionals and enhance the client's right to demand access to services. Scientists and health-care professionals suggested that such a system was not appropriate in the British context, worrying that such a practice would flood their system with demanding patients while obscuring access for those who could benefit most. One molecular geneticist described the problem in the following manner. "And what we need to, there has been quite a strong emphasis on this in this country on trying to develop, I don't think it's in place yet, but trying to develop a system that gives equitable access to these services, but gives also some kind of gateway function. And the gateway can operate both ways, really, it can operate as a funnel into access to something, but it's also a controlling function. And I do think that if you had completely open access, it wouldn't be a good use of either public or private resources."[53] Triage systems, which were pivotal to the maintenance of a government-run system with scarce resources, had emerged in tandem with an authoritative health-care professional who controlled access to health care. If triage was abandoned in the case of BRCA testing, patients could become more demanding of genetic medicine and in health care more generally, perhaps destabilizing the health-care professional's role and the NHS's investment in genetics services in the future.

Opponents were particularly worried that Myriad's service would hurt the growing infrastructure of genetic medicine in the NHS. If genetic tests were offered privately, neither molecular nor clinical geneticists within the NHS would have access to the increasingly common and complicated cases in genetic medicine. The CMGS noted: "Removal of significant income streams, removal of key elements of the analytical process and exclusion from experience with developing technologies will impoverish Regional Genetics Centres and cause a stagnation and loss of morale that is hard to reverse."[54] The BRCA genes were the longest and most complex genes that had been found to date. Without developing the infrastructure to test for these genes, how would health-care professionals deal with the next complicated gene that was bound to come along, for heart disease or obesity? The CMGS also worried that offering BRCA testing privately just as genetic medicine was poised to become more integral to mainstream medicine would jeopardize the NHS's role in the provision of genetic testing: "At best, UK centres would be deskilled to the level of sub-contractors of Myriad Genetics for routine work.... A feature of Clinical Molecular Genetics in the last 10 years has been the rapidity of transferring research funding to tests of clear benefit to patients. Unless Regional Centres and the research groups with whom they collaborate are exposed to the problems of applying leading edge technologies to diagnostics, development will be increasingly confined to commercial companies."[55] The CMGS also pointed out the instability of commercial testing, arguing that a company "may be bought out, go out of business, or decide to end a particular area of testing for commercial or other reasons."[56] Citing Oncormed's demise as an example, it suggested that a sophisticated genetics infrastructure within the NHS was pivotal to ensure the continued and uninterrupted availability of genetic testing. While this movement of genetic testing from research laboratories to the commercial sector had taken place relatively seamlessly in the United States, this development was extremely controversial in Britain. Not only did it jeopardize the expertise and infrastructure that had been developed by genetics professionals in the NHS, but it also challenged a cornerstone of the NHS's major preventive efforts.

Would Myriad's expensive and demand-based testing system interfere with the British commitment to provide all clients equal access to health care? Some critics argued that if testing was available on demand, it would be provided in an uneven manner across the country. Access would be based on initiative and financial opportunity, rather than a family history of breast and ovarian cancer. The CMGS report stated that such a demand-based system could have very damaging consequences for the NHS: "On the one hand it threatens . . . spiralling costs and on the other hand geographic inequalities of access to diagnosis."[57] Within a health-care system that promised health care equally based on need rather than income, such a result could undermine the legitimacy of the system.

Indiscriminate BRCA testing, many health-care professionals argued, could also devastate the NHS mission to provide clients with health care based on need by drastically limiting the number of clients who could be tested. If the NHS had to pay the high costs of Myriad's full DNA sequence analysis (about $2,500 per test) within its limited budget, they would be able to offer the test to far fewer clients than the current system allowed. One geneticist stated that Myriad's test was simply too expensive to fulfill the NHS's commitment to effectively allocate resources to maintain the public's health: "Because we've got a month's delay to send the DNA abroad, we are paying over the odds for cost, and the whole principle of the NHS is that it should be cost-neutral."[58] Rationing schemes would have to become stricter, and many clients with extensive family histories of breast and/or ovarian cancer might not qualify for BRCA testing. Neva Haites, who had initially been more open to a relationship between Myriad and European health-care providers, said: "In a way I'd rather offer a 70 percent service to the whole of the UK rather than a 100 percent service to a tenth of the country."[59] She suggested that while their laboratory techniques might not pick up all mutations, combining it with risk-assessment services allowed for it to be offered more inexpensively, to more people, and with clinical goals in mind. Wendy Watson also questioned the rationing implications of using Myriad's service: "It's, will genetic testing become more rationed than it should be because of the extra expense. And if that happens, I shall fight it. That's where I am coming from. I'm not particularly bothered if somebody's patented a part of my gene or whatever. That's not the issue. The issue is that it might reduce the number of people who are able to have genetic testing, who may well die because they haven't had genetic testing. And that is wrong."[60] While this statement seems to contradict Watson's strong opposition to gene patenting quoted earlier in this chapter, it actually points us to the main concern of Myriad's critics. Scientists, health-care professionals, and even Wendy Watson were less concerned with the patenting of genes themselves than on the implications that such a practice would have for the NHS, which had considerable public support.

In addition to questioning the validity of patent rights that were used as a justification of Myriad's European expansion, British health-care professionals and scientists could not accept a testing system built on the model of the independent diagnostic laboratory. They argued that it was not valid in the British context both because it did not require risk assessment or counseling and because full-sequence analysis would not help to achieve their primary goal: to manage women at increased risk of breast and/or ovarian cancer. They also challenged the roles for system participants—health-care professionals without gatekeeping authority and an client with a right to demand laboratory analysis—that Myriad's service prescribed.

Resolution

The UK Department of Health, British scientists, health-care professionals, and patient advocates negotiated with Myriad for more than a year, trying to reach an agreement that would be acceptable to all parties involved. By late 1999, however, it had become clear that opposition to Myriad was intractable and widespread—neither British health-care professionals nor clients would be likely to welcome the company. Some health-care professionals even threatened to bring Myriad to court if the company tried to enforce its patents. In addition, there were indications that Mike Stratton might sue Myriad for illegally acquiring a license to his patent. (The scientist's license to Oncormed had explicitly disallowed Myriad's owner-ship.) Myriad, however, remained persistent. Britain, after all, could be the gateway to a potential gold mine of European patients.

The National Health Service was particularly concerned that the esca-lating controversy and acrimonious environment might lead Myriad to demand increased royalties or licensing fees as part of a future agreement with the Department of Health or even proceed with litigation. Dr. Sheila Adam, the Health Services Director for the NHS, tried to address the wide-spread opposition of health-care professionals by issuing a letter to "col-leagues" at regional genetics centres, requesting their cooperation to work with the NHS as it negotiated with Myriad. Her letter expressed worries about the possibility of litigation, noting that the patent situation was not clear:

As a condition of any license granted, CRCT [Stratton] has required the licence-holder to grant to NHS a free sub-license to practice BRCA 2-testing. However, there is currently no granted UK patent on BRCA 1, although DH [Department of Health] has been advised by the UK Patent Office that the Myriad patent applications are expected to be granted shortly. Department solicitors advise that the NHS labora-tories which offer BRCA 1 testing are potentially infringing Myriad Genetics' intel-lectual property rights. If Myriad Genetics is successful in gaining patent protection in the UK, . . . [it] could choose to take action against NHS laboratories, claiming damages back to the date on which the patent claim was filed (August 1996). It is our understanding that these damages could be substantial.[61]

During this period, Myriad began to negotiate with private laboratories to offer BRCA testing to the UK population, and in March 2000 the company announced that it had issued a license to Rosgen Ltd., an Edinburgh-based private genetics laboratory, which would offer laboratory analysis of the BRCA genes—within a context of pre- and post-test counseling—on a fee-for-service basis.[62] Patients with private health insurance or who could afford to pay out of pocket could utilize the faster and arguably more tech-nically sensitive services. At the time, however, this agreement did not affect Myriad's ongoing negotiations with the NHS.

The architecture of the testing system created by the Myriad-Rosgen agreement represents a fascinating compromise between the US and British systems for two reasons. First, Myriad appeared to learn from its battle and allowed Rosgen to incorporate counseling into its testing system—thereby creating a service that was more in keeping with the priorities of British health-care professionals and the NHS. Dr. Pete Kitchin, Managing

Director of Rosgen, said: "Our aim is to ensure that such NHS patients in the United Kingdom are able to gain the widest possible access to the best possible testing. However, we must stress that Rosgen will offer the test only if appropriate pre-test and post-test counseling services are in place."[63] Although Rosgen did not specify what type of counseling should take place, Myriad had retreated from its original position and created a testing system that looked different from its American version. Second, while Rosgen seemed to accept the British approach to counseling, it adopted Myriad's definition of a client who had a right to choose testing. When it stated in its promotional materials that it would offer the widest possible access, Rosgen implicitly argued that it would provide testing to those clients whose lack of risk factors denied them access to NHS services. Although British health-care professionals had emphasized the ideal of providing everyone with equal access to the system, Rosgen adopted Myriad's approach to provide the "best possible testing" to the widest audience possible.

Despite the compromise represented by the Myriad-Rosgen service, British health-care professionals were reluctant to direct patients to it and continued to use the National Health Service's BRCA-testing system. In June 2000, when Rosgen sent letters announcing its service to general practitioners across Britain, staff members of the Southwest Thames Regional Genetics Service issued a vehement response:

Much of the original work in mapping the genes for BRCA1 and BRCA2 was done on families in the South West London and Surrey area as part of the charitably funded work by the Cancer Research Campaign at the Institute of Cancer Research and The Royal Marsden Hospital. This work was put in the public domain and Myriad Genetics has claimed a patent for BRCA2 on the basis of sequencing the remainder of the gene. It seems ironic therefore that relatives of these individuals who gave samples to improve medical science for all should potentially have testing prejudiced by this commercial interest.[64]

These genetics professionals still found the Myriad-Rosgen system problematic, because it had not resolved questions about the patentability of and ownership rights over genes or the implications of a genetic testing monopoly for the future of research and health care. Such frustrations were not inconsequential. In a country with very limited direct-to-consumer marketing of medical products, Rosgen relied on these professionals to provide referrals.

In November 2000, Myriad and Rosgen reached an agreement with the UK Department of Health. The settlement allowed the National Health Service to continue testing without paying royalties or licensing fees to Myriad. In what was hailed as an unprecedented deal, Myriad and Rosgen agreed to waive royalties on all breast cancer genetic tests that had or would be provided by the National Health Service while Rosgen agreed to provide the NHS with data about the mutations it collected in order to improve the NHS's clinical services.[65] Rosgen would continue to provide testing privately in the United Kingdom, to those clients who could afford the fee of £179–2,600 (depending on which test was performed). This deal represented another major concession by Myriad. By agreeing not to interfere with NHS's testing service, it had lost the majority of revenue and gene sequence information from BRCA-testing services in Britain. The company, it seemed, had agreed to a competitive environment of multiple testing-system architectures similar to the one it had successfully dismantled in the United States.

The fate of the Myriad-Rosgen-DoH deal, however, was threatened in January 2001 when Rosgen filed for voluntary liquidation for reasons unrelated to their agreement with the NHS. Rosgen's collapse meant that Myriad no longer had a presence in Britain. Because its deal with the NHS had been based on its license with Rosgen, Myriad could choose to renegotiate, though it had not done so as of fall 2006. Considering recent developments in Europe, which will be discussed in further detail in the Epilogue, it is not likely that the company will try to re-enter Britain.

Conclusion

When Myriad attempted to expand its BRCA-testing service by exerting its patent rights in Britain, it encountered a very different terrain than the one it successfully dominated in the United States. In the United States, a few "cease and desist" letters to competitors and lawsuits sufficed to establish a monopoly. In light of this easy domination at home, and in light of the fact that the British testing system did not involve full sequence BRCA testing, one might expect that transferring Myriad's technology would be easy. But Myriad's efforts in Britain were unsuccessful because the values and approaches to biomedicine that were embedded in the company's testing system conflicted directly with the elements of the British toolkit that developers in that country had used to build its testing systems.

Using patent rights, Myriad tried to transfer not just a full-sequence test but an entire social, political, and economic system based on an independent diagnostic laboratory, in which the laboratory limited a client's access to testing only by ability to pay, did not require counseling, and imposed few restrictions on the physician who ordered the test. In Britain, however, scientists and patient advocates, among others, explicitly rejected the patentability of disease genes and the use of such patents to create commercial genetic-testing services. In addition, while Americans had become accustomed to a commercialized genetic testing environment by the time Myriad built its system, the British continued to offer all of these services within the NHS and with an integrated approach to counseling and DNA analysis—a model that it had also followed in the development of BRCA testing. Furthermore, Myriad's system seemed to challenge long-held views about the rights and the roles of physicians and patients in Britain. While Myriad's testing system identified the same phenomenon—mutations to the BRCA genes—as the NHS system, the approach to biomedicine embedded in the company's system could not work in Britain as it had in the United States. Myriad's system was made up not only of technical details, but also deeply embedded national norms, values, and approaches to biomedicine that added additional challenges to the attempt to transfer its technology across national boundaries.

Conclusion

James Watson, co-discoverer of the structure of the DNA molecule, once famously remarked: "We used to think our future was in the stars. Now we know our future is in the genes." Watson was wrong. Our genomic futures are by no means preordained. Throughout this book, we have seen that the way genetics is understood, genetic technologies developed, and genetic medicine built is being fundamentally shaped by national social and political context.

In both the United States and Britain, distinctly national toolkits of ideas—assembled because of specific regulatory frameworks, laws, histories, and traditions—shaped how genetic testing for breast cancer was envisioned, constructed, and stabilized. In the United States, the multiple systems that were built to test the BRCA genes for mutations reflected the history of genetics services in the country. Over the previous decades, genetic-testing services had grown with the expansion of the biotechnology industry, and the government had been reluctant to step in and standardize their provision or use, citing both lack of resources and a desire to support the growth of biotechnological innovation. The government had also demonstrated its support of the biotechnology industry by ensuring not only that strong patent protection was available, but also that universities could patent and commercialize research findings that were the result of federal funding; these decisions would prove to be important to the development of genetic medicine and BRCA testing in particular. In the United States, a market environment for genetic medicine has been built with a variety of providers—including academic medical centers, reproductive-services clinics, and private laboratories—offering services in multiple forms, such as DNA analysis with and without specialized counseling. Within this market environment, four major BRCA-testing services were built. The Genetics and IVF Institute, a private reproductive-services clinic, was the first to offer BRCA testing purely as a commercial service, but did so on a very limited scale by only testing for the three mutations common among the Ashkenazi Jewish population. Myriad also built its technology, DNA analysis that was available through any physician for a fee, as a consumer product. It offered a variety of laboratory analyses which made it useful for a wide variety of clients concerned about their BRCA risk. Overall, its testing system focused on providing "state-of-the-art" laboratory services, allowing clients to use its test with or without specialized counseling. Two providers of BRCA testing did, however, try to heed the recommendations of expert advisory committees, as well as the positions articulated by scientific and medical organizations and patient advocacy groups. Oncormed offered BRCA testing in the context of clinical research protocols (while still making its product available to as many consumers as possible), and the University of Pennsylvania's Genetic Diagnostic Laboratory offered DNA analysis only through specialized clinics at academic medical centers.

This variety of BRCA-testing services in the United States did not last long. Adopting familiar American strategies of litigation and marketing, Myriad soon became the sole provider of BRCA testing in the United States. It shut down its rival testing services by threatening patent infringement, a strategy common among patent holders, particularly those in the American biotechnology industry. As it threatened litigation against the other providers of BRCA testing, however, Myriad also created new definitions of research and health care. Only those tests in which results were not returned to clients could be classified as "research" and were therefore outside Myriad's purview as patent holder. Every other system qualified as "health care" and thus required a license from the company covering use of the BRCA genes for clinical purposes. This strict distinction would have long-term consequences. Not only would it limit availability of the test, but many critics argued that Myriad's approach to its patent rights prevented the testing process from being properly refined, which would have serious implications for clients. In fact, in an article published in the *Journal of the American Medical Association* in March 2006, researchers at the University of Washington discovered that the company's testing process missed a large number of deleterious mutations. This was evidence, they suggested, that Myriad's test was flawed, and much more research was necessary—research that was very difficult to conduct within the company's constraints.[1]

Simply eliminating the other providers of the test through strict control over its patent rights, however, wasn't sufficient. In order to ensure that people purchased its test, Myriad tried to convince health-care professionals and clients that its technology would be empowering and would enhance their individual autonomy. The company thus tapped into age-old American themes of individualism and empowerment in order to market its product. The company characterized its testing system as a contribution to a long history of patient and women's empowerment, while also appealing to health-care professionals' desires to be independent consultants rather than beholden to insurers, lawyers, hospitals, the government, and even bioethicists.

In Britain, BRCA testing was also built in the image of the other genetics services that had been developed before it. Testing has been developed through the National Health Service, with regional genetics clinics, which offered laboratory analysis in the context of counseling, as the primary sites of care. For these British providers, standardizing the clinical dimensions (rather than the laboratory dimensions) of BRCA testing—including where care was provided, what type of care was offered, and who had access to services—was most important. In contrast, most American services focused their BRCA-testing technologies on the activities of the laboratory.

There was, however, disagreement in Britain over how best to interpret the NHS's goal of providing equal access to care in the development of BRCA testing, particularly given finite resources. Early developers of BRCA testing offered multiple solutions to this problem, cognizant both of the high potential demand for the new technology and the likelihood that BRCA testing would be seen as a model for the future of genetic medicine. Some developers suggested that NHS's regional genetics clinics were best equipped to determine how testing services should be provided to their populations, as they knew how to build systems that would best serve local populations. Others suggested the creation of a national system that would couple a strategy of risk assessment and triage that was quite familiar within the NHS with an initiative to identify both moderate-risk and high-risk individuals. Proponents of this national BRCA-testing system eventually encouraged acceptance of their approach by launching a comprehensive educational and marketing campaign across the country to convince health-care professionals that their strategy could best ensure national equity and care, efficient use of scarce resources, and provide a rational model for the provision of genetic medicine that could ensure future funding of these services within the NHS. They used a consensus-building approach, which was common not only for the development of clinical-practice guidelines within the NHS, but also used broadly throughout British policymaking. In Britain, as in the United States, developers used elements familiar to their national context—risk assessment and triage, attention to providing citizens with equal access to health care, focus on prevention and public health efforts—to both build and market the new technology.

The architectures of the BRCA-testing systems that eventually dominated in the United States and Britain were quite different from one another. Myriad focused its attention on providing DNA analysis, allowing clients to choose how they wanted access to a test, whether through a specialist geneticist, primary-care physician, oncologist, surgeon, or other type of physician. Within the national BRCA-testing system in Britain, in contrast, the NHS and regional genetics clinics maintained control over both clinical and laboratory aspects of the system. Not only did British providers of BRCA testing require that DNA analysis be accompanied by counseling, but they spent considerable time working to standardize the clinical elements of the technology—risk assessment, triage, and counseling. While DNA analysis was the focus of Myriad's system, the activities of the clinic were considered most important in the British system.

These testing systems also served different purposes, reflecting very different understandings of a good health outcome. While Myriad, as a private company, focused on encouraging use of its test and finding clients with BRCA mutations, the NHS, influenced by its broad public health and prevention objectives, sought to identify all those at elevated risk for breast and/or ovarian cancer. Such distinct priorities led providers in the two countries to characterize the role of BRCA testing in predicting risk and disease and guiding prevention efforts in very different ways. Trying to demonstrate the utility of its laboratory technology, Myriad created categories of both risk and disease according to the presence or absence of a BRCA mutation, even emphasizing the existence of a new disease of inherited cancer susceptibility for which the main symptom was BRCA-mutation-positive status. It also publicized the availability of tamoxifen to cure the disease that its technology could diagnose. Indeed, the use of the drug as a chemopreventive could also justify the widespread availability of the company's testing service. Meanwhile, the British national BRCA-testing system classified and managed clients according to risk categories that were defined by family history, including a moderate-risk category. Proponents of this system argued that they needed to manage all those at elevated risk for breast and/or ovarian cancer, not just those with BRCA mutations who might have a disorder of cancer susceptibility. Such broad public health concerns also explain their reluctance to prescribe tamoxifen as a chemopreventive; NHS health-care professionals and officials had to worry about the long-term health of all British women.

The different architectures of BRCA testing in the United States and Britain, in terms of whether and how the test was marketed, how access to it was obtained, what type of counseling and laboratory analysis was available, and what was done with the results, shaped the rights, roles, and responsibilities of providers, health-care professionals, and clients of BRCA testing quite differently in the two countries. The GDL's decision to offer its laboratory analysis only through genetics clinics at academic medical centers restricted the client's access to testing but did not interfere with the autonomy of the genetics professionals who offered clinical care. Meanwhile, Oncormed's strategy to test only high-risk clients enrolled in research protocols also restricted clients' access but also placed additional burdens on health-care professionals who had to provide counseling according to standardized guidelines and help clients get access to research protocols.

Ultimately, the tests that were successful in dominating their national environments defined their users in rather different ways. Myriad characterized its client as an empowered consumer, who could choose the type of clinical care she wanted to receive in conjunction with laboratory analysis and who eventually could use genomic information to make informed medical decisions. Both health-care professionals and the company itself were defined as facilitators, helping consumers use BRCA testing to understand and deal with their cancer risk. Control over BRCA-testing services was distributed quite differently in the national system run by the British National Health Service. Through the national model of risk assessment and triage initially devised by the UK Cancer Family Study Group and the Public Health Genetics Unit and later validated by the National Institute of Clinical Excellence, NHS officials and health-care professionals decided who could gain access to which components of the testing system. They were gatekeepers, adopting the traditional role of the medical establishment. Clients in this system were both traditional patients and citizens—they were forced to heed their physician's advice, but were guaranteed equal access to health care and did not have to worry about affording genetic testing because they were members of the tax-paying public. The availability of tamoxifen in the two countries also prescribed similar identities for the participants in health care. In the United States, researchers, AstraZeneca, Myriad, and the Food and Drug Administration made the drug available and allowed consumers to choose whether they wanted to use it; in Britain, health-care professionals, the National Health Service, and the National Institute of Clinical Excellence decided that it was too premature to offer tamoxifen as a chemopreventive and protected their citizen-patients from a potentially dangerous drug.

This book has shown that national contexts play important roles in shaping scientific and technological development. Developers draw on distinct toolkits as they build the architectures of their innovations. In addition, the architectural differences that we have observed are not merely unimportant technicalities; they have important implications for users—in this case, the identities of participants in health care, definitions of risk, disease, and appropriate preventive measures, and even the processes of international technology transfer.

Diverging Futures of Genetic Testing

The story of the development of BRCA testing told throughout this book suggests that the field of genetic medicine is moving in rather different directions in the United States and Britain. In the United States, a commercial environment has emerged, with consumers able to gain access to DNA analysis through a variety of service providers. In fact, recent events suggest that it is becoming even easier to get access to genetic-testing services without the involvement of a physician. In 2005, DNA Direct, a privately held company based in San Francisco, began marketing the use of an online interface to allow the public to gain access to genetic-testing services directly. The company offered online access to a number of genetic tests, including three of the BRACAnalysis tests sold by Myriad (it does not provide access to Myriad's Rapid BRACAnalysis).[2] A client interested in using one of the BRCA tests through this interface visits DNA Direct's website to order a test, and then calls one of the company's experts to discuss her personal and family history. If she chooses to pursue DNA analysis, the client goes to a patient service center managed by LabCorp, where she pays and her blood is drawn. LabCorp sends the payment and blood sample to Myriad's laboratories for analysis. (As was discussed in chapter 3, Myriad and LabCorp had an ongoing partnership, working together to market BRCA testing widely.[3]) After three to four weeks, the client receives an e-mail stating that the test results are available, calls DNA Direct, and receives her results over the telephone from one of DNA Direct's genetic counselors. These tests cost approximately $200 more per test than Myriad's services (ranging from $585 to $3,311), because they include both Myriad's DNA analysis and DNA Direct's telephone consultations. The process for BRCA testing, however, is slightly different than for the other genetic tests offered by DNA Direct. In these other cases, including testing for the cystic fibrosis and hemochromatosis genes, a kit is sent directly to the home.[4] This kit includes directions, a cheek swab, and forms for payment and a signed informed consent form. After swabbing her cheek, the client returns this sample, along with payment and informed consent, to DNA Direct in a postage-paid return envelope. No clinical interaction, not even the telephone consultation required for the BRCA-testing service, is required. Results of the test are available online within 7–10 days. Presumably, DNA Direct built a slightly different system for BRCA testing because of its relative newness, high visibility, and controversial past.

DNA Direct has built a suite of genetic testing systems that does not require the involvement of an independent health-care professional. Instead, the company relies on its website and the telephone and e-mail advice of its genetic counselors to inform clients about the meaning and implications of the tests. Clients of DNA Direct's system seem to have complete control over whether and how they want to use genetic-testing services, and do not have to engage with an independent physician at all. Indeed, while Myriad envisions physicians as simply playing a facilitative role, in DNA Direct's systems the doctor-patient relationship has been eliminated; the central relationship envisioned by this technology is between the testing company and the client. The architecture of these testing systems resembles that created for the at-home blood-pressure kit. In the case of at-home blood-pressure testing, however, the FDA influences the technological architectures considerably, in terms of what types of information must be provided with the test and how it should be provided, as well as its safety and efficacy.

Despite this increasing commercialization, discussions about how best to provide genetic susceptibility testing continue. The US Centers for Disease Control and Prevention (CDC) are trying to develop recommendations to guide the clinical interaction that is part of the genetic testing process. Also, as recently as September 2005, the US Preventive Services Task Force, an independent expert panel convened by the government's Agency for Healthcare Research and Quality, revisited the controversy over the appropriate provision of BRCA testing and recommended that a client should use BRCA testing only if her family history of breast and/or ovarian cancer had been found to be associated with a BRCA mutation.[5] Of course, CDC's efforts and the approach suggested by the Task Force are the same as the one taken by the British national BRCA testing system. Unlike PHGU and NICE in Britain, however, the CDC and Task Force have very limited influence on the provision of BRCA testing services in the US; they can only hope that individual health-care professionals and clients will heed their recommendations.

In Britain, meanwhile, genetic testing continues to be provided almost entirely through the NHS's system of regional genetics clinics, with counseling and testing offered in an integrated manner. There have been, however, isolated attempts to create private genetics services. Medichecks, a private laboratory, has begun to offer genetic testing via the internet (for a fee of £555, approximately $1000.)[6] Its testing systems appear quite similar to DNA Direct's, with mail-order test kits and a nurse advisor providing the client with results and relevant post-test counseling over the phone. While such services stimulated the government's Human Genetics Commission (HGC) to write a report, the government has yet to pass any legislation on the issue.[7]

The diverging infrastructures for genetic testing in the United States and Britain also suggests different futures for the users of health care in these countries. In many respects, the empowered consumer defined by the architectures of Myriad's and DNA Direct's systems seem to be triumphs of late-twentieth-century patient advocacy and bioethics movements to eliminate a paternalistic approach among physicians and help patients take charge of their own health care. In comparison, the approach taken in the British national system seems much more traditional. Physicians determine the course of care for their patients, who can refuse access to care but cannot demand it. This book has demonstrated, however, that empowerment has multiple meanings and that initiatives to increase patient empowerment can bring both benefits and costs. While the consumer of Myriad's BRCA-testing system has virtually unlimited autonomy to choose to take BRACAnalysis and decide which management options are right for her, she is also saddled with the responsibility of making these decisions. As American patients are increasingly defined as consumers empowered to make decisions about their health care, they are also expected to take on many of the burdens previously left in the domain of the health-care professional, including informing themselves about the risks and benefits of medical interventions, sorting through complicated risk statistics and deciding whether to take a test or a drug that promises both risks and benefits. In contrast, although patients using the British BRCA-testing system certainly have less control in determining the course of their own health care, they are more clearly and strongly supported by health-care professionals and the NHS who take clear responsibility for their welfare. Although empowerment certainly seems like wonderful progress for the users of health care, particularly in a country that valorizes individual entrepreneurship as much as the United States does, it is important to consider its meanings and positive and negative implications as we encourage this approach to the provision and use of health care.

Beyond Breast Cancer: Informing Technology Policymaking

Genetic medicine is still in its early stages in the United States and Britain, and its shape has not completely stabilized. As more genetic tests are offered, their provision continues to be a subject of considerable public debate. Genetic testing for Alzheimer's Disease is now available, but it raises the same kinds of issues as BRCA testing, as the genes found are linked only to a small subset of patients who contract the disease at an early age. Related technologies are also being developed that remind us of our concerns about genetic testing while also creating new dilemmas. Pharmacogenetic testing, which some hope will be used to identify DNA markers (which themselves do not cause any disease) that make individuals receptive to particular treatment regimens, has stimulated debate about whether it is possible or desirable to create therapies targeted to the individual and if such efforts will re-inscribe race as a relevant biological category.[8] Another similar breed of genetics, called nutrigenomics, has already emerged, which promises to identify genetic variants linked to particular body types and health needs and thereby help clients make better health and lifestyle decisions. Sciona, Inc. offers a nutrigenetic test called Cellf to discover "The Science of You." The Cellf Report, which is sent to the client after she buys a kit at a retail outlet and sends a cheek swab and payment to the company, explains "which gene variations you have, what effects they may have on your health, and what specific amounts of food, nutrients, supplements, and exercise may help—tailor-made and written just for you."[9] Pre-implantation genetic diagnosis, which allows couples to genetically test an embryo created through in vitro fertilization for a variety of conditions, raises questions not only about safety and utility but also about whether selecting an embryo for implantation on the basis of genetic-test results leads, through a "backdoor," to eugenics.[10] All these tests have again elicited calls for the regulation of genetic testing, and, in particular, for an initial investigational period before such products are made commercially available. This growth in genetic medicine continues to be accompanied by vigorous discussions among scientists, physicians, patient advocates, private companies, and scholars, over how these technologies ought to regulated, provided, and used, how they should be owned and commercialized, and how to ensure that an individual's genetic privacy is maintained.

The comparative case study that has been conducted throughout this book contributes to current conversations about the appropriate directions of genetic medicine in four ways.

First, by providing an in-depth look at how genetic medicine is being built, including why and how specific technological choices are made and what the consequences of these choices are, we can make better and more informed decisions about the types of regulatory frameworks we might want to devise for this emerging field. Its comparative approach also provides us with multiple alternatives for the regulation and provision of genetic technologies, and allows us to assess the benefits and risks of each policy choice. For example, this book has explored the consequences of offering widespread genetic testing almost immediately after genes are discovered. When is a technology ready for widespread use? Is it more important to make technologies available quickly or to ensure that they are safe and effective before they are offered widely? Is there a viable compromise in between? As was noted earlier, the March 2006 *JAMA* article suggests that Myriad's test could not detect a significant proportion of BRCA1 and BRCA2 mutations.[11] Might the creation of an intense investigational period before widespread use, or a requirement to couple DNA analysis with genetic counseling, reduce the likelihood and impact of false negative results? Might an investigational period also increase the amount of epidemiological data available, regarding the relationship between gene mutations and disease incidence, to improve the utility of tests? Of course, in Britain these issues are considered by NHS and NICE, whereas in the United States the FDA takes on this responsibility—although it has chosen not to deal with such questions in the case of genetic testing. This book has demonstrated, however, that not only do these issues arise again and again in the development of genetic medicine, but that the way these issues are resolved has serious consequences for the public. Myriad's choice to offer BRACAnalysis soon after gene discovery meant that there were considerable uncertainties generated by test results, regarding the meanings of positive and negative results and how to interpret the presence of a variant of uncertain significance. In addition, in both the United States and Britain, testing was made widely available before the direct utility of medical management options were clear. We now have case-study information to supplement the warnings of the Task Force on Genetic Testing and the Institute of Medicine in the early 1990s. This information can facilitate our discussions of how best to enact regulatory frameworks for genetic testing.

Second, the book demonstrates how particular laws, policies, values, and norms can shape technological development in fundamental ways. We often assume that the process of innovation is a closed and objective one, with developers following a linear and clearly determined path. What we have seen throughout this book, however, is that the way to make the "best" technology varies widely, and is decided differently by developers depending on their moral, social, political, and economic context. This finding suggests that policymakers have an important role to play in shaping technologies throughout the developmental process. Technology policymaking has traditionally been restricted to two domains: laws that facilitate innovation through direct funding or the creation of an environment receptive to commercialization, and regulatory frameworks that shape the provision and use of a technology once it has already been built. This approach often leads to severe restrictions at early stages of research (e.g., cloning, embryonic stem cells) or limiting use of a technology once it has already been built (e.g., nuclear power.) We have seen throughout this comparative case study, however, that decisions about what a technology means and how it should be provided and used are made throughout the developmental process. For example, the decision by British proponents of the national BRCA-testing system to restrict access to BRCA testing through risk assessment and to families in which an affected member could be tested first not only privileged the importance of family history but also suggested that the new technology was simply an additional tool to be integrated into the services available, rather than an entirely new technology that could have clear implications for risk, disease, and prevention.

If policymakers are concerned about and want to shape the implications of technologies, particularly in domains subject to massive public concern, then they must do so in a nuanced way, after detailed assessments, and early—at "upstream" moments in the process. Upstream technology assessment has also been advocated by other scholars of science and technology policy, in order to "reduce the human costs of trial and error learning in society's handling of new technologies, and to do so by anticipating potential impacts and feeding these insights back into decision making, and into actors' strategies."[12] In fact, once a technology has been built, it is often too difficult to enact policies to govern its provision and use because so many such decisions have already been implicitly made and are therefore embedded in the architecture of the technology. Rather than conceptualizing the process of innovation as a linear path that can either be enabled or disabled, policymakers should see it as a process that can take multiple paths, depending on the judgments of those involved in the developmental process. Such an upstream effort can also benefit innovators. Rather than being confronted with public controversy or policies that could significantly influence uptake of a technology after many years have been spent on the developmental process, policy deliberation and action at early stages in the innovation process could lead not only to more socially desirable outcomes but also technologies that are more profitable.

Third, this book provides tools to conduct the kinds of assessments needed to intervene in the kind of technology policymaking described above. By conceptualizing technologies in terms of their architectures, we can understand not only the components and how they are fitted together, but also understand how each element of a technology's architecture is both connected to its context and has consequences for users. How Myriad and British proponents of the national standard chose to make their technologies available shaped the rights and responsibilities of the health-care professionals and clients who used their technologies. By understanding the implications of technologies in terms of their components, we can determine how a technology needs to be built in order to influence its social consequences. Of course, one cannot fully anticipate all the consequences of a technology's architecture, but in-depth investigations of similar technologies and prospective assessments at early stages of development can help us identify the most likely scenarios, the potential positive and negative consequences, and how best to build a technology to maximize known benefits and minimize known risks. Furthermore, this book demonstrates how comparative technology assessment both within and between countries can be useful, allowing the analyst to highlight the implications of similarities and differences in both the developmental process and the way a technology is built. It also provides insight into the alternative architectures and paths of technology development available. Among the four BRCA-testing systems that initially emerged in the United States, is there one that we prefer? Would genetic testing in the United States benefit from regulatory intervention that requires the provision of specialized counseling alongside laboratory analysis? Should the United States create an analogue to Britain's NICE, to standardize the care of clinicians and acknowledge that they are pivotal parts of technological systems? Should the British encourage the expansion of genetic testing in the private domain, to allow clients to be empowered to choose to use whatever technologies they wish? Of course, as we consider these alternative frameworks, we should also remember that certain elements of these genetic testing technologies will be very difficult to transport, as they are not only firmly connected to national toolkits, but contradictory to elements in other contexts. It would be quite difficult to institute a national system of risk assessment and triage in the United States, and as we have already seen, it would be quite unlikely for the British NHS to construct a testing system that did not involve counseling from regional genetics clinics.

Finally, this book has demonstrated how the inscription of national context into technological architectures can create significant challenges for economic globalization. Indeed, when Myriad tried to expand its testing system to Britain, it was not simply making its sophisticated laboratory technology available in a country with limited access to genetics services. It was trying to impose specific approaches—to the commercialization of science and medicine, the doctor-patient relationship, public health, and prevention—on a country that had already articulated very different, and often opposing, priorities in the development of genetic medicine. It is not at all surprising, then, that Myriad was met with tremendous resistance from British scientists, health-care professionals, patients, and government officials. The vigorous controversy that erupted suggests that rather than serving as bridges of globalization through their objective and universal nature, technologies can serve as embodiments of particular national norms, values, and traditions and become flashpoints for transnational conflict. We have already seen other episodes of transnational technological conflict—consider recent disputes about genetically modified organisms between the United States and European Union, and the uneasy adoption of Western (and mostly American) approaches to intellectual property in the developing world—and they are likely to occur more frequently particularly as countries knit themselves together to create a global economic future. If we begin to acknowledge that social, political, and economic approaches are embedded in the scientific findings and technological developments that we are attempting to transfer, however, we may be able to anticipate and even mitigate, or at least better understand, the transnational challenges we may face.

Epilogue

Recent events in Europe suggest that the British opposition to Myriad's attempted technology transfer was not isolated, but rather is one of a series of episodes of growing discontent with the globalization of American approaches to health care and intellectual property. As expected, in 2001 and 2002 the European Patent Office granted a total of five patents on the BRCA genes. A patent covering the BRCA2 gene (and all diagnostic uses) was issued to Mike Stratton and the Cancer Research Campaign (now called Cancer Research UK). Myriad was issued four patents, three covering the normal sequence, mutations, and diagnostic and therapeutic uses of BRCA1 and one covering a number of mutations to BRCA2. Almost immediately after Myriad's patents were issued, a broad coalition of groups—representing scientists, health-care professionals, governments, patients, and environmentalists—took advantage of the European Patent Office's opposition mechanism and challenged all of the company's patents. (See table E.1.) They did not mount a similar challenge to Stratton's BRCA2 patent, because Stratton promised to license it freely throughout Europe.[1] These groups, which included scientists who had been invited to tour the company's laboratory in 1998, had witnessed Myriad's attempt to shut down the British National Health Service's BRCA testing system and expected that the company would use the same strategy throughout the rest of Europe. Their efforts were even supported by the European Parliament, which passed a resolution challenging the legality of Myriad's BRCA patents and their consequences for the European public in 2001.[2]

The groups challenged Myriad's patents on ostensibly technical grounds (e.g., lack of priority and absence of novelty, lack of inventive step, insufficient description of the invention, and lack of industrial applicability) but as we have seen throughout this book, these technical disputes were

Table E.1

Challengers to Myriad's patents at the European Patent Office.

Scientific organizations	Italian Society for Human Genetics German Society for Human Genetics Danish Society for Medical Genetics Swiss Society of Medical Genetics National Center for Scientific Research "Demokritos" (Greece) Society of Medical Genetics (Czech Republic) Austrian Society for Human Genetics British Society for Human Genetics Finnish Society for Human Genetics Belgian Society of Human Genetics European Society of Human Genetics Institut Curie
Health-care Professionals	Belgian Centres for Human Genetics Dutch Society of Clinical Genetic Centres Swiss Cancer Research Institute Institut Gustave-Roussy Assitance-Publique-Hôpitaux de Paris Fédération Nationale des Centres de Lutte Contre le Cancer (French national federation of anti-cancer centers) Fédération Hospitalière de France (French federation of hospitals)
Governments	Dutch Ministry of Health Austrian Ministry of Health Belgian Ministries of Health, Social Affairs, and Scientific Research European Parliament
Patient groups	Borstkanker Vereniging Nederland (Dutch breast cancer advocacy group) Vlaamse Liga tegen Kanker (Belgian cancer advocacy group) Deutsche Krebshilfe (German cancer league, a cancer charity and support group)
Other NGO	Greenpeace Germany
Political parties	Swiss Social Democrat Party

simultaneously social, economic, and political ones. As they and their lawyers questioned the accuracy and the novelty of both the gene sequences and diagnostics patented by Myriad, these groups clearly were questioning the company's approaches to science and health care and proposing their own. Their rhetoric echoed the initial British criticisms, as they questioned the company's inventiveness, worried about the future of human genetics research in Europe, and suggested that Myriad's patents violated a European approach to public health.

In documents and presentations that accompanied their challenges at the European Patent Office, Myriad's opponents reminded people of the international effort to find the BRCA genes in the 1980s and the early 1990s. Although it had been characterized as a "race," they suggested that all of the research, including incremental findings such as linkage analyses and discoveries of genetic markers, contributed to the final mapping and sequencing by Myriad and Stratton's group. The BRCA gene discoveries were, they argued, the result of a long collective effort. In a press release explaining its position, Institut Curie—the biomedical research institution that spearheaded much of the EPO opposition—argued as follows:

Myriad Genetics may have won the very last stretch in the race to breast and ovarian cancer predisposition genes in 1994, but between 1990 and 1994 the international public consortium had singly achieved detailed localization of the BRCA1 gene, and provided significance as to its features, and its possible use in the detection of breast and ovarian cancer susceptibility. What remained to be done was the final gene sequencing, a technological procedure the outcomes of which warrant, at the most, protection by limited monopoly rights.[3]

The final gene-sequencing effort undertaken by Myriad, critics argued, did not require inventiveness but rather just time and technological infrastructure and was thus unpatentable. Representatives of the Dutch and Belgian genetics societies, the other major groups that took a significant role in the EPO opposition, conceded: "Evidently, the identification of BRCA1 was a laborious and costly effort. However, neither this nor the reduction of the time to reach an obvious goal, involves 'inventive step' in the sense of patent law."[4] These critics emphasized the collective and integrated dimensions of scientific investigation, and suggested, as their British counterparts had, that it simply didn't make sense to either carve up the research into discrete patentable parts or reward the group that mapped and sequenced a gene first.

European opponents' second major line of argument was that Myriad's BRCA patents would hinder innovation, which could have consequences for the European biomedical community and the public. The European Parliament's resolution articulated a fear that Myriad's patents "could seriously impede the development of and research into new methods of diagnosis."[5] Others described these worries in more detail. A representative from the Belgian genetics society worried that a Myriad monopoly would "cost the European medical community expertise. . . . Myriad says anyone is allowed to use the sequence for research, but no one is interested in sequencing patient samples unless you can return the result to the patient or publish in a journal."[6] Like the British opponents of Myriad, these groups worried that patent restrictions would prevent European scientists from doing research that was necessary for both their careers and the welfare of the public.

This issue was of particular concern for European geneticists because they had already seen how the BRCA gene patents could stifle biomedical research and consequently diminish the quality of medical care. In 2001, Dominique Stoppa-Lyonnet (a scientist at Institut Curie) and her colleagues reported that with a new DNA analysis technique they were able to find a new BRCA1 mutation, a large rearrangement in the gene, that had not been detected by Myriad's DNA-sequencing techniques.[7] Mutations due to such large genomic rearrangements were said to account for at least 10 percent of all BRCA mutations. Although Myriad had previously criticized BRCA-testing services in Europe as a "hodge-podge of mostly low sensitivity inaccurate testing,"[8] European scientists charged that the new evidence about genomic rearrangements demonstrated how Myriad's testing system itself was inaccurate and therefore had negative implications for clients. Myriad seemed to eventually accept this criticism, as it later supplemented its analytic methods to test for these large rearrangements.

European opponents argued that discovery of the rearrangements clearly demonstrated the need for continued development of genetic testing techniques, for the immediate benefit of clients with BRCA mutations as well as the long-term development of genomic medicine: "Comparative analysis of the different methods . . . taking both sensitivity and cost into consideration, are now needed to improve genetic testing for breast and ovarian cancer predisposition."[9] Such research, however, would have been prevented by Myriad's exertion of its patent rights. Stoppa-Lyonnet and her colleagues had returned the results of their analysis to the client, a practice that Myriad considered to be a "clinical service" rather than "research" and thus an infringement of its patents. It is unlikely, as discussed throughout this book, that clients would enroll in a research protocol in which they did not receive test results, particularly when BRCA-testing services were widely available. Opponents argued that Myriad's patent restrictions would prevent researchers from developing and refining genetic testing techniques and thus jeopardize the future of genetic medicine.

And opponents argued that Myriad's patents could jeopardize distinctively European approaches to health care. A number of groups worried about the costs of Myriad's services, arguing that state-run health systems would be forced to reduce access to testing or allocate more funding to genetics.[10] In addition, both Institut Curie and Greenpeace Germany (another opponent of the EPO patents) suggested that Myriad's separation of testing and counseling conflicted with the combined risk assessment, counseling, and laboratory analysis system that had emerged in France, Britain, and much of Europe.[11] "This approach," Institut Curie explained, "goes very much against the way we view public health care, in France and in most other European countries, where clinicians work within a model which integrates biological research, clinical investigation, and patient care, taking into account the medical and psychological aspects of diagnosis as well as the clinical history of high risk patients and their families."[12] A system like Myriad's, they argued, would be bad for clients particularly in light of the uncertainties about the meaning of BRCA mutations for disease incidence and risk management. As these groups registered their frustrations and opposition to Myriad, they also articulated a specifically European approach to health care. Although there were differences in the funding and provision of health care between European countries, the opposition to Myriad—both in terms of the groups involved and the rhetoric they used—defined a common European approach to health care and a common right of the European citizen which were distinct from its American counterparts.

These challenges to Myriad's patents have been successful. In decisions in 2004 and 2005, the European Patent Office forced the company to narrow its patents so much that the company will likely be unable to interfere with existing European BRCA-testing services except in one respect: the company still holds a patent on one BRCA2 mutation that is found in the Ashkenazi Jewish population. This means that under current law, in order for any European laboratory to test individuals for this mutation, it must either send the sample to Myriad or get a license from the company (if allowed to do so). This decision has caused some controversy. Critics charge that this rule could lead to discrimination—that members of this ethnic group will be forced to buy Myriad's expensive testing service, because they will not have access to BRCA testing through one of the European state-run systems. Gert Matthijs of the Belgian Society of Human Genetics articulated these concerns: "Women coming to be tested for breast cancer will have to be asked whether they are Ashkenazi Jewish or not. If they are, the health-care providers will only be able to offer the test if they paid for a license, or they will have to send the women's samples abroad. Women who are not Ashkenazi Jewish–or who just don't know that they have Ashkenazi Jewish ancestors–will be entitled to a test which is free. This is the first time that this kind of situation has arisen in genetic testing, and we find it very worrying."[13] Here too, as in the initial development of BRCA testing in the United States and Britain, we see how patent decisions can shape not only technological development, but have significant consequences for users. This decision, however, can still be appealed. In addition, it is still unclear how Stratton's BRCA2 sequence patent may interfere with Myriad's narrowed patent on one BRCA2 mutation; depending on Myriad's continued ambitions in Europe, the rights of these patent holders may be determined in the courts.

Meanwhile, Myriad is still trying to set up European outposts of its testing services. In Germany it has licensed its patents to Bioscientia, one of the largest clinical testing laboratories in Europe.[14] In order for clients to use these services, physicians need to send a blood sample, along with payment and family-history information, to the laboratory, who will return results within a few weeks. Although it is unlikely that Myriad will be able to use its patent covering the BRCA2 gene to eliminate other BRCA-testing services in Europe, it continues to work hard to develop a presence on the continent. European opposition to the company, however, is not likely to change.

Methodological Appendix

This book is based on a variety of sources gathered from 1998 to 2004, including published articles in newspapers, magazines, and scientific journals, semi-structured interviews, documents generated by those involved in the development of BRCA testing in the United States and Britain, and ethnographic observation. In this appendix, I explain the logic behind the use of these sources and describe, in some detail, how I gathered this information.

Published Sources

I began my research on this project in the spring of 1998 by searching the internet, the general media, and scientific journals for information about the BRCA gene discoveries and the development of BRCA testing. Gathering this information had three purposes. First, it helped me sketch out a basic understanding of the chronology of the story. Second, I was able to identify the major topics of controversy, and follow how these debates progressed through published sources. Finally, I identified the major stakeholders in these discussions, which allowed me to assemble a list of the people I wanted to interview and the questions I wanted to ask them. This list included primarily the American and British participants in the stories being told (e.g., the scientists who discovered the BRCA genes, officials at the institutions developing testing), health-care professionals at genetics clinics providing access to testing, the bioethicists and activists who commented on the development of the new technology, and even the journalists and authors who had followed the gene discoveries and development of the technology before me.

As the project progressed, I periodically updated my libraries of published documents, in order to stay abreast of developments in BRCA research and testing as well as ongoing discussions about how the technology was being provided and used and its implications for the future of genetic medicine. This attention proved particularly important; as I was doing my research and writing the manuscript, the story evolved considerably—Myriad's attempted technology transfer was challenged first in Britain and then throughout Europe, and studies published in the 2000s suggested that Myriad's "gold-standard" test was not picking up all deleterious mutations.

Interviews

Overall, I interviewed 111 individuals in the United States, Britain, Belgium, Germany, and The Netherlands. I began with the list of stakeholders I had generated through my preliminary research, and then identified additional interviewees using a snowball sampling method in order to ensure that I had spoken with everyone involved in the development of genetic testing for breast cancer in the United States and Britain. In the snowball sampling method, the interviewer chooses a group of initial interviewees based on a set of specific criteria (as stated earlier, I looked for individuals identified in the media or other similar sources as participants in or commentators on the development of genetic testing for breast cancer), asks each interviewee about other individuals that she should interview because they might have information pertaining to the study, interviews the individuals who have been referred and asks them for further contacts, and continues this process until the entire network of individuals who might be important participants to include in the study has been identified and interviewed. (See Bijker 1995.) For example, I asked the actors that I interviewed during my first interview trip to Britain about other individuals in the United States and Britain with whom I should speak, interviewed the people suggested by the first group of interviewees and also asked them about other individuals that I should interview, and continued in this manner until I had exhausted the pool of individuals whom I had identified or had been suggested by interviewees as being relevant to the story. This interviewing method was useful not only because it allowed me to identify and interview a large group of individuals, but also because I was able to understand the social and political networks that emerged in tandem with the new technology. In fact, identifying the details of these networks led me beyond my two main countries of study to debates that were occurring at the European Commission and European Patent Office. In sum, I interviewed the following types of individuals in both countries:

providers of each BRCA-testing system

patient advocacy group representatives

medical charity representatives

representatives of other non-governmental organizations

representatives of insurance industry associations

representatives of biotechnology industry associations

government officials involved in the provision or use of genetic testing

European Commission officials

patent office officials

members of both governmental and non-governmental advisory committees who deliberated on issues related to genetic testing

scientists involved in the BRCA gene discoveries

genetic counselors

primary-care physicians and specialists offering access to BRCA testing

scholars involved in discussions about the appropriate provision or use of genetic testing.

Each of my interviews was semi-structured. In advance of each interview, I developed a guide to help direct the conversation. These guides did not articulate the exact questions that I would ask or the order in which I would raise particular issues, but rather provided an outline of themes that I wanted to cover during the course of the interview. This technique helped me ensure that I covered all issues of particular interest to me while also providing flexibility to tailor the interview according to the responses of the interviewee. It also allowed me to conduct the interview in the form of a conversation; This was particularly important, I thought, because it might elicit more genuine responses and diminish the awkwardness of a formal interview. While each interview guide varied according to the individual I was interviewing, most focused on the individual's role in the development of genetic testing and their perception of broader issues, including the provision and regulation of genetic testing and genetic information. I took notes during these interviews and also taped them in order to facilitate their future transcription and analysis.

Each of these interviewees signed an informed consent form that described the purpose of the study and how I planned to use the interview data. In the consent form, I stated that I would do my best to maintain the interviewee's anonymity. While it is sometimes difficult to believe that this type of social science research has any potential risks, I recognized that in the highly politicized environment of genetic testing, subjects might be less likely to speak freely if they thought their statements could be easily traced back to them. In the consent form I also noted that if I found it impossible to use an interviewee's statements without making his or her identity obvious, I would seek additional permissions before publishing anything. This occurred only once. In that case, I told the subject which statements I planned to include and asked if it was acceptable to attribute them to him or her. If the subject did not agree, I simply did not include that subject's statements in the manuscript.

Although I interviewed similar types of individuals in both countries, I found that my interviews in Britain were somewhat more useful than those I conducted in the United States. Some individuals and institutions in the United States simply refused to be interviewed. Others, including some individuals who were in charge of government advisory bodies, were reluctant to go beyond their official policy statements and recommendations to comment on how the statements were developed. There may be at least two reasons for this national difference. First, the politics of BRCA testing were much more adversarial in the United States, which may have led interviewees to be much more careful about their use of language. Second, corporate representatives may have been worried about disclosing proprietary information. In fact, this was the reason given when I was denied access to most top officials at Myriad Genetics. While I was able to interview one top-level official at the company and a genetic counselor, most of their staff refused to speak with me, despite repeated requests via e-mail, telephone, letter, and even in person. I dealt with these interview difficulties by acquiring information about the company through both archival work and ethnographic observation.

Document Analysis

While interviews provided me with a great deal of information, I also relied heavily on documents produced by stakeholders that tangibly described how BRCA testing should be or was already being regulated, provided, and used. I was particularly interested in gathering the following types of documents:

reports by government and other advisory committees who issued recommendations on topics relevant to the development of BRCA testing

position papers and policy statements from advocacy groups and scientific and medical organizations regarding BRCA testing

investor information for commercial providers of BRCA testing, including annual reports, filings to the Securities and Exchange Commission (SEC), and assessments of the company by stock analysts

press releases issued by stakeholders involved in the development of the new technology

promotional and educational information about BRCA testing, from testing providers and others

documents that accompanied the testing system itself, including requisition forms, family history forms, consent forms, and sample test results.

I gathered many of these documents from the individuals that I interviewed; I asked clinicians at genetics clinics to provide me with all of the documentation that they had on their genetic testing services, from promotional materials to informed consent forms. I also found some policy statements related to my topic online (from, e.g., the American Medical Association, the British Society for Human Genetics, and the American College of Obstetrics and Gynecology). In some cases, I simply called organizations (e.g., the National Society of Genetics Counselors) and asked them to send me any policy statements or position papers related to my areas of interest.

In a few cases, I had to collect as many documents as possible in order to make up for the insufficiency of information gathered through interviews. My attempts to gather information about Myriad Genetics provide a perfect example of this. How could I research a topic that involved Myriad as one of the major players and not speak with the director of its laboratory, its chief executive officer, or its chief scientific officer? I dealt with this dilemma by gathering as many of the publicly available documents generated by the company as possible. A great deal of information was available through the company's website, and I also looked through the documents that it had filed with the SEC. I also received publicity and educational materials, as well as test results forms (with all identifying information redacted) from scientists and health-care professionals that I interviewed.

Ethnographic Observation

It was also very important to me to observe the controversies over BRCA testing in action. In order to do this, I attended a number of conferences at which the new technology was to be discussed, including meetings of the American Society of Human Genetics (San Francisco and Philadelphia), public forums on genetic testing (London, Chicago, and Los Angeles), the UK Forum for Genetics and Insurance (London), the US Secretary's Advisory Committee on Genetic Testing (Washington), and the World Conference on Breast Cancer Advocacy (Brussels). I also attended EPO hearings on the BRCA1 gene in Munich. I took copious notes of all of these meetings; I taped and transcribed parts of them as well.

Notes

Introduction

1. Meyer et al. 2003. For more on national policy differences in approaches to science and technology, see Brickman et al. 1985; Gottweis 2002; Willems et al. 2000. (Works cited by name and year are listed in the bibliography.)

2. Brickman et al. 1995.

3. President Bill Clinton, speech on completion of first survey of entire human genome, June 26, 2000.

4. Annan 2003.

5. Rabinow 1999.

6. Árnason and Simpson 2003.

7. For more information, see the following webpages: Facts about Breast Cancer in the United States: Year 2002 (at http://www.natlbcc.org); Breast Cancer Statistics (at http://www.ukbcc.org.uk); Breast Cancer: The Facts (at http://www.breastcancercampaign.org).

8. Singletary and Kroll 1996; Love and Lindsey 2000.

9. Ragaz et al. 1997; Lerner 1999; Leopold 1999.

10. National Breast Cancer Coalition, "Position Statement on Screening Mammography," press release, March 2003; Olsen and Gøtzsche 2001; Miller et al. 2002; Nyström et al. 2002.

11. Lerner 2001; Leopold 1999; Lorde 1980; Kushner 1982, 1985, 1975, 1980; Friedan 2001, 1993.

12. Stabiner 1998; Casamayou 2001.

13. Mosca et al. 2000.

14. Davies and White 1996.

15. Hall et al. 1990.

16. Miki et al. 1994.

17. Nigel Hawkes, "Scientists identify breast cancer genes," *Times* (London), September 15, 1994; "Breast cancer gene isolated," *Gazette* (Montreal), September 15, 1994; "Gene that causes breast cancer isolated," Agence France Presse, September 14, 1994; "Mutant gene found tied to breast cancer," Xinhua News Agency, September 15, 1994.

18. Natalie Angier, "Scientists identify a mutant gene tied to hereditary breast cancer," *New York Times*, September 15, 1994. The news was reported on the front pages of a number of newspapers across the world. See, for example, Richard Saltus, "Mutated gene tied to early breast cancer is located," *Boston Globe*, September 15; "Big step against breast cancer," *Houston Chronicle*, September 15; Thomas H. Maugh II, "Discovery of breast cancer gene called major advance," *Los Angeles Times*, September 15, 1994; Tim Friend, "Inherited breast cancer gene located," *USA Today*, September 15; David Perlman, "Hereditary breast cancer gene found," *San Francisco Chronicle*, September 15; David Brown, "Gene for an inherited form of breast cancer is located," *Washington Post*, September 15.

19. Tim Friend, "Inherited breast cancer gene located," *USA Today*, September 15, 1994.

20. Wooster et al. 1995.

21. Clive Cookson, "UK team wins race to find second breast cancer gene," *Financial Times*, December 21, 1995.

22. Easton 1994; Easton et al. 1995; Struewing et al. 1997; Fodor et al. 1998; Thorlacius et al. 1998.

23. Nelkin and Lindee 1996; Hedgecoe 1998.

24. Lippman 1991.

25. Zola 1972; Fox 1977; Chang and Christakis 2002; Clarke et al. 2003.

26. Duster 2003; Lock 1998.

27. Wexler 1992.

28. Holtzman and Shapiro 1998.

29. Ellen Goodman, "The breast-cancer gene: Finding it only leads to tough ethical questions," *Pittsburgh Post-Gazette*, September 21, 1994.

30. Vogel 1986.

31. Brickman et al. 1985; Brickman 1982.

32. Jasanoff 1991.

33. Jasanoff 1995.

34. Patel and Pavitt 1994. For more on the National Innovation Systems approach, see Edquist 1997; Metcalfe 1995.

35. Jasanoff 2005.

36. Ibid.

37. Bijker et al. 1989.

38. Latour 1988.

39. Casper and Clarke 1995.

40. Bijker 1997.

41. Ibid.

42. Hughes 1983.

43. Ibid.

44. Ibid.

45. Skocpol 1997.

46. Swidler 1986.

47. Clinton, speech, June 26, 2000.

48. von Beuzekom 2001; "Origin of US biotechnology patents," *Chemical and Engineering News* 79, no. 44 (2001): 56.

49. Bijker et al. 1987; MacKenzie and Wajcman 1999.

50. Bijker et al. 1987; MacKenzie and Wajcman 1985.

51. Kline and Pinch 1996.

52. Latour 1987; Callon 1987; Casper and Clarke 1995.

53. Wooster et al. 1995; Ford et al. 1998; Tavtigian et al. 1996.

54. Woolgar 1991.

55. Akrich 1992, p. 208.

56. Oudshoorn and Pinch 2003.

57. Wailoo 1997, p. 2.

58. Lock 1998.

Chapter 1

1. National Research Council 2002.

2. Starr 1984.

3. Paul 1995; Kevles 1998.

4. Lindee 2000.

5. President's Commission for the Study of Ethical Problems in Medicine and Biomedical and Behavioral Research, "Screening and Counseling for Genetic Conditions: The Ethical, Social, and Legal Implications of Genetic Screening, Counseling, and Education Programs," 1983.

6. Ibid.

7. Wright 1994.

8. Starr 1984; Abbott 1988.

9. Thomas et al. 2002; Cho et al. 2003.

10. National Society of Genetic Counselors, "Taking a Stand." Downloaded February 19, 1999 from http://www.nsgc.org.

11. Clinical Laboratory Improvement Act of 1988, Public Law 100–578, October 31, 1988.

12. Hilgartner 2000.

13. Institute of Medicine 1994.

14. US Food and Drug Administration, Center for Devices and Radiological Health, "Device Classes." Downloaded September 6, 2005 from http://www.fda.gov.

15. Andrews et al. 1994.

16. Ibid.

17. Task Force on Genetic Testing, "Promoting Safe and Effective Genetic Testing in the United States" (National Human Genome Research Institute, 1997).

18. Ibid.

19. Ibid.

20. Ibid.

21. US Congress, House of Representatives, Committee on Science, Subcommittee on Technology, "Technological Advances in Genetics Testing: Implications for the Future" (hearing, September 17, 1996).

22. Edward R. McCabe (Chair, Secretary's Advisory Committee on Genetic Testing) to Donna E. Shalala (Secretary of Health and Human Services), letter, April 24, 2000:

"On behalf of the Secretary's Advisory Committee on Genetic Testing (SACGT), I am writing to express our support. . . ."

23. Starr 1984.

24. Klein 2001.

25. UK Sociologist No. 1, interview by the author, August 1, 2005.

26. Webster 1998.

27. R. Robinson and J. LeGrand, "Evaluating the NHS Reforms" (King's Fund Institute, 1993); UK Geneticist No. 4, interview by the author, October 7, 1999.

28. Greg Dyke, "The New NHS Charter—A Different Approach" (London: Department of Health, 1998).

29. Some private obstetrics and gynecological clinics emerged, but as of the mid 1990s no private genetics clinics or laboratories had been built (Pollock 2004).

30. Coventry and Pickstone 1999.

31. Department of Health and Social Security, "Prevention and Health, Everybody's Business: A Reassessment of Public and Personal Health" (Her Majesty's Stationery Office, 1976).

32. Ibid.

33. Ibid.

34. Ibid.

35. Ashmore et al. 1989.

36. Department of Health and Social Security, "Prevention and Health, Reducing the Risk: Safer Pregnancy and Childbirth" (Her Majesty's Stationery Office, 1977).

37. Human Genetics Advisory Commission, "The Implications of Genetic Testing for Insurance" (London: Department of Trade and Industry, 1997).

38. Human Genetics Advisory Commission, "The Implications of Genetic Testing for Employment" (London: Department of Trade and Industry, 1999); Human Genetics Advisory Commission and The Human Fertilisation and Embryology Authority, "Cloning Issues in Reproduction, Science, and Medicine" (London: Department of Trade and Industry, 1998).

39. Advisory Committee on Genetic Testing, "First Annual Report: July 1996–December 1997" (London: Health Departments of the United Kingdom, 1998).

40. Advisory Committee on Genetic Testing, "Code of Practice and Guidance on Human Genetic Testing Services Supplied Direct to the Public" (London: Health Departments of the United Kingdom, 1997).

41. Advisory Committee on Genetic Testing, "Report on Genetic Testing for Late Onset Disorders" (London: Health Departments of the United Kingdom, 1998).

42. Ibid.

Chapter 2

1. Natalie Angier, "Scientists identify a mutant gene tied to hereditary breast cancer," *New York Times*, September 15, 1994.

2. Natalie Angier, "Fierce competition marked fervid race for cancer gene," *New York Times*, September 20, 1994.

3. Member of the National Breast Cancer Coalition, interview by the author, December 8, 1999.

4. Breast Cancer Action, "Policy on Genetic Testing for Breast Cancer Susceptibility, USA," 1996. In an article titled "From the Executive Director: The future is now" (*Breast Cancer Action Newsletter* 40, 1997), Barbara Brenner wrote: "Genetic counseling requires a multidisciplinary approach, involving expertise not only in genetics but in psychological issues as well."

5. National Breast Cancer Coalition, "Genetic Testing for Heritable Breast Cancer Risk," 1996.

6. Ibid.

7. The NBCC articulated three primary objectives: to promote research, improve access, and increase its influence ("History, Goals, Accomplishments," 2000).

8. NBCC, "Genetic Testing for Heritable Breast Cancer Risk."

9. Brenner, "From the Executive Director: The future is now."

10. Breast Cancer Action, "BCA's policy on genetic testing for breast cancer susceptibility," *Breast Cancer Action Newsletter*, June 1996.

11. Ibid.

12. Barbara Brenner, "Off with their breasts!" *Breast Cancer Action Newsletter* 2, 1996.

13. National Breast Cancer Coalition, "Commentary on the ASCO Statement on Genetic Testing for Cancer Susceptibility," 1996.

14. Epstein 1996.

15. Ibid.

16. Boston Women's Health Book Collective 1973.

17. Ruzek 1978.

18. Breast Cancer Action, "Taking Action, Creating Change: Breast Cancer Action Annual Report 1997."

19. Brenner, "Off with their breasts!"

20. Gieryn 1983.

21. Pinn and Jackson 1996.

22. American Society of Clinical Oncology, "Statement of the American Society of Clinical Oncology: Genetic Testing for Cancer Susceptibility 10," 1996.

23. American Society of Human Genetics 1994.

24. American Society of Clinical Oncology, "Statement of the American Society of Clinical Oncology: Genetic Testing for Cancer Susceptibility 2," 1996.

25. Ibid.

26. The University of Pennsylvania was, in fact, one of the leading academic medical centers in the US—in addition to being considered one of the country's best hospitals, the university's academic medical center ranked very high in monetary value of federal grants received (University of Pennsylvania press releases: "The Penn School of Medicine Receives Record $327 Million from NIH in FY 2001," March 22, 2002; "Hospital of the University of Pennsylvania Garners Kudos From U.S. News & World Report's Annual Rankings," July 12, 2002.

27. Ganguly et al. 1993.; Williams et al. 1995.

28. US Geneticist No. 3, interview by the author, March 21, 2000.

29. For more information on genetic counseling, see Hsia et al. 1979.

30. Their recommendations were based on a variety of statistical models that had been developed to assess risk. The most common were the "Gail" and "Claus" models (Gail et al. 1989; Claus et al. 1994).

31. US Geneticist No. 3, interview by the author, March 21, 2000.

32. Ibid.

33. Star and Griesemer (1989) write: "Boundary objects are objects which are both plastic enough to adapt to local needs and the constraints of the several parties employing them, yet robust enough to maintain a common identity across sites." According to Star and Griesemer's classification of boundary object types, the forms

in GDL's system would be considered "standardized forms": "boundary objects devised as methods of common communication across dispersed work groups."

34. US Geneticist No. 3, interview by the author, March 21, 2000.

35. OncorMed Inc., Annual Report, 1995.

36. OncorMed Inc., Annual Report, 1994.

37. Kenney 1998; Kleinman 2003.

38. Thomas et al. 2002.

39. Anderson 1991.

40. Eisenberg 2002.

41. Cancer Research Campaign and OncorMed, "Code of Practice for Companies Seeking a Licence or Sub-Licence to the Diagnostic Rights of the BRCA2 Patent," 1997.

42. US Geneticist No. 2, interview by the author, January 11, 2000.

43. "Federal policy for the protection of human subjects: Notices and rules," *Federal Register* 56 (June 18, 1991): 28002–28032. The June 1991 *Federal Register* announcement is the only publication of this "Common Rule" that governs the appropriate conduct of human subjects research.

44. "To help ensure that the highest ethical and medical standards are followed when its genetic predisposition testing services are offered to patients, the Company employs institutional review board-approved protocols and clinical testing guidelines . . ." (OncorMed Annual Report, 1995).

45. OncorMed Annual Report, 1995.

46. Committee on Labor and Human Resources, 104th Congress, second session, "Advances in Genetics Research and Technologies: Challenges for Public Policy," 1996.

47. Office of the Inspector General, US Department of Health and Human Services, "Institutional Review Boards: The Emergence of Independent Boards," 1998.

48. The guidelines required that health-care professionals gather certain information from the individual, including the subject's age, sex, ethnicity, and family history of breast or other cancers and outline the type of information that the health care professional should provide to the individual interested in testing. They also required that health care professionals cover the following topics in the counseling session: information about the purpose of the test; the individual's option to be tested; benefits and risks of the test; sensitivity and relevance of possible results; clinical implications and limitations of results; possible psychological stress; who

will have access to the results; relevance of the results to relatives and how to communicate such results to them; potential effects on health and life insurance policies; what will happen with the person's DNA sample; that consent is required before data is released to third parties; circumstances of release of non-BRCA2 data, including gender of fetus and misattributed paternity. Source: Cancer Research Campaign and OncorMed, "Code of Practice for Companies Seeking a Licence or Sub-Licence to the Diagnostic Rights of the BRCA2 Patent," 1997.

49. Protein truncation testing detects nonsense and frame-shift mutations (Stratton et al. 1999).

50. Genetics and IVF Institute (http://www.givf.com).

51. Gina Kolata, "Breaking ranks, lab offers test to assess risk of breast cancer," *New York Times*, April 1, 1996.

52. Genetics and IVF Institute, advertisement for BRCA testing services, *Los Angeles Jewish Times*, September 6, 1996.

53. Barbara Koenig, "Gene tests: What you know can hurt me," *New York Times*, April 6, 1996; Hubbard and Lewontin 1996; Burke et al. 1996.

54. Hadassah, "Policy Statement on Genetic Testing," Chicago, July 1997.

55. Schulman and Stern 1996.

56. Rothman 2003.

57. Schulman and Stern 1996.

58. Duster 1990; Nelkin and Tancredi 1991; Rothstein 1999.

59. Schulman and Stern 1996.

60. Recent scholarship in the field of science and technology studies, however, has highlighted the difficulties of defining human groups, and particularly, how defining the makeup of human groups takes place on a highly contested scientific and political terrain. See Reardon 2001.

61. Genzyme homepage (http://www.genzyme.com), retrieved January 20, 2003.

62. Duster 1990.

63. Myriad Genetics annual report, 1995.

64. Myriad Genetic Counselor, interview by the author, April 3, 2000. Myriad employed people trained as genetics counselors to handle a variety of public education tasks, such as responding to enquiries from physicians and individuals about its testing services, discussing the testing service at national conferences of the American Society of Human Genetics and National Society of Genetic Counselors, and working with activists to allay their fears about testing.

65. Myriad Genetics, Inc., ". . . I did something new today to guard against cancer" (advertisement, *New York Times Magazine*, September 1999).

66. Myriad Genetics, Inc., "Myriad Genetics Launches Direct to Consumer Advertising for Breast Cancer Test," press release, September 12, 2002. The company also gave interested individuals information about how to access specialized genetic counseling centers through its website. The website included a searchable database of "Risk Assessment Centers" nationwide that offered "genetic counseling and testing services to individuals at increased risk for hereditary breast and ovarian cancer." Myriad Genetics Inc., Search for Genetic Counseling and Testing Centers, 1999 (http://www.myriad.com, visited September 8, 2000). While the company did not want to become involved in determining the type of counseling that should be provided or the health care professional who would be in charge, it did provide individuals with the opportunity to access such services if they so desired.

67. American Society of Human Genetics, "ASHG Response to Request for Public Comments on Preliminary Final Recommendations on Oversight of Genetic Testing," 2000.

68. These prices increased over time. While Comprehensive BRACAnalysis initially cost $2,400, for example, it cost $2,600 in 2001. In addition, some states added charges to testing. In New York state, for example, BRACAnalysis cost $2,680 in 2000.

69. Leeson 1979.

70. Wendy Watson, interview by the author, 1998.

71. Bryan Christie, "Genetic research gave clue to risk of breast cancer," *Scotsman*, May 29, 1996.

72. Wendy Watson, interview by the author, 1998.

73. National Breast Cancer Coalition, "Legislative Accomplishments." Downloaded September 7, 2005 from http://www.natlbcc.org.

74. Genetic Interest Group, "The Present Organisation of Genetic Services in the United Kingdom," June 1995.

75. Coventry and Pickstone 1999.

76. The amount of funding that a regional genetics clinic received from the NHS to pay for BRCA-testing services depended on numerous factors, including the source from which the clinic was requesting the funding, the population of the region, and the existing expertise and infrastructure of the clinic. For example, regional genetics clinics used a variety of NHS sources to fund their efforts. Some clinics which had been involved in research to find the breast cancer genes reallocated this money to the development of BRCA-testing services. Others requested

additional funds from the regional NHS purchasers that were in charge of funding health care services for a particular region. A few appealed to the research and development section of the central NHS, proposing that the development of BRCA testing should be treated as an investigational venture rather than one with clear clinical benefits. In some cases, regional genetics clinics contracted with other laboratories to conduct the analysis of the BRCA genes. (UK Geneticist No. 1, interview by the author, August 10, 1999.)

77. Although private health insurance was widely available in the UK by the 1990s, no genetics services were available through private physicians. When some individuals chose to send their blood to Myriad's laboratories in the US for analysis, private insurance companies did not reimburse them for these costs.

78. Clinical Molecular Genetics Society, "Familial Breast Cancer." Downloaded January 20, 2003 from http://www.cmgs.org.

79. British Society of Human Genetics, "What Happens at a Medical Genetics Appointment?" Downloaded October 15, 2002 from http://www.bshg.org.

80. UK Geneticist No. 4, interview by the author, October 7, 1999.

81. It would be inaccurate to suggest that the British were unconcerned by issues of genetic privacy and discrimination, as they needed to purchase private life insurance in order to get a mortgage on their home. The way these issues were addressed and incorporated into policymaking, however, differed considerably from the United States. See Parthasarathy 2004.

82. This effort to develop a national strategy for specialized services was not unique to BRCA testing. During 1998 and 1999, the NHS developed a consultation document for review by providers of specialized services across the country, in an attempt to "renew the NHS as a genuinely national service. Patients will get fair access to consistently high quality, prompt, and accessible services right across the country" (NHS Executive, "Draft Guidance: Commissioning Specialized Services," 1998).

83. UK Geneticist No. 4, interview by the author, 1999.

84. Ibid.

85. Celia Hall, "Cancer reform aims at equality of care," *Independent* (London), May 16, 1994, p. 4; Celia Hall, "Cancer patients promised best care," *Independent* (London), May 19, 1994, p. 5; Lois Rogers, "UK cancer care worst in Europe doctors say," *Sunday Times* (London), April 3, 1994.

86. UK Geneticist No. 4, interview by the author, October 7, 1999.

87. Working Group for the Chief Medical Officer, London, UK Department of Health, Genetics and Cancer Services, 1996.

88. Ibid.

89. Ibid.

90. Ibid.

91. The classification scheme also took into account the ethnic background of the individual interested in testing. If an individual was of Ashkenazi Jewish descent, she was usually automatically placed into the high risk category and referred to the genetic clinic for counseling and testing.

92. UK Public Health Genetics Unit representative, interview by the author, July 19, 1999.

93. Gail et al. 1989; Benichou et al. 1996; Claus et al. 1994.

94. Angier, "Scientists identify a mutant gene."

95. James Mackay et al., "Familial Breast Cancer: Managing the Risk" (Anglia Clinical Audit and Effectiveness Team, 1997).

96. Mackay and Zimmern conducted a number of regional training seminars to ensure that these interactions would be standardized.

97. Some regions did not use the exact risk thresholds suggested by Mackay and Zimmern. While the exact risk thresholds may have varied slightly among regions, however, all regions used some type of risk assessment, classification, and triage scheme (R&D Office of the Anglia and Oxford NHS Executive and Unit for Public Health Genetics, "Report of Consensus Meeting on the Management of Women with a Family History of Breast Cancer," 1998).

Chapter 3

1. OncorMed also negotiated and received licenses for the BRCA1 and BRCA2 genes from Mary-Claire King at the University of Washington and from the UK Cancer Research Campaign. Myriad had filed for patents on its own research on BRCA1 and BRCA2, while OncorMed supplemented patent applications on its own research (BRCA1 and BRCA2) with exclusive licenses on patents from Mary-Claire King's localization work and Mike Stratton's BRCA2 work. Myriad ran into problems regarding propriety when developing its intellectual-property portfolio. NIH Director Harold Varmus threatened to file for a counterpatent on the BRCA1 gene when the company's initial patent application for the gene did not mention NIH members of the Myriad team as co-inventors. They eventually settled the issue when NIH was provided with some license fees, the NIH researchers were listed as co-inventors on the patent, and the NIH abandoned its own patent application. The patent that was eventually accepted, then, was owned solely by Myriad. See "Dispute arises for patent over gene," *New York Times*, October 30, 1994.

2. Patricia D. Murphy, Antonette C. Allen, Christopher P. Alvares, Brenda S. Critz, Sheri J. Olson, Denise B. Schelter, and Bin Zeng, "Consensus sequence of the human BRCA1 gene," US Patent 5654155, issued August 5, 1997.

3. Donna M. Shattuck-Eidens, Jacques Simard, Francine Durocher, Mitsuuru Emi, Yusuke Nakamura, "Linked breast and ovarian cancer susceptibility gene," US Patent 5693473, issued December 2, 1997.

4. Donna M. Shattuck-Eidens, Jacques Simard, Francine Durocher, Mitsuuru Emi, Yusuke Nakamura, "Linked breast and ovarian cancer susceptibility gene," US Patent 5709999, issued January 20, 1998; Mark H. Skolnick, David E. Goldgar, Yoshio Miki, Jeff Swenson, Alexander Kamb, Keith D. Harshman, Donna M. Shattuck-Eidens, Sean V. Tavtigan, Roger E. Wisement, and Andrew P. Futrea, "17q-linked breast and ovarian cancer susceptibility gene," US Patent 5710001, issued January 20, 1998.

5. US Geneticist No. 2, interview by the author, January 11, 2000.

6. Mary-Claire King, Lori Friedman, Beth Ostermeyer, Sarah Rowell, Eric Lynch, Csilla Szabo, and Ming Lee, "Genetic markers for breast, ovarian, and prostatic cancer," US Patent 5622829, issued April 22, 1997; Phillip Andrew Futreal, Richard Francis Wooster, Alan Ashworth, Michael Rudolf Stratton, "Materials and methods relating to the identification and sequencing of the BRCA2 cancer susceptibility gene and uses thereof," US Patent 6045997, issued April 4, 2000.

7. Myriad Genetics, Inc., "Myriad Genetics Obtains OncorMed's BRCA1/BRCA2 Genetic Testing Program in Patent Settlement," 1998.

8. Peter Meldrum, "Myriad Genetics, Inc.," *Wall Street Transcript*, December 1998.

9. William A. Hockett, Myriad Genetics, Inc., to unidentified individual at University of Pennsylvania, date unknown: "I understand that you are currently providing diagnostic testing services for BRCA1."

10. For more on this tension, see Löwy 1997.

11. US Geneticist No. 3, interview by the author, March 21, 2000; Christopher L. Wight, Myriad Genetics, Inc. to Robert R. Terrell, University of Pennsylvania, September 22, 1999: "I acknowledge receipt of your letter of September 10, 1999, and appreciate your willingness to discuss and resolve this matter with Myriad."

12. Myriad did allow a few laboratories to test some samples for mutations. The Diagnostic Molecular Genetics Laboratory at Memorial Sloan-Kettering Cancer Center (New York) and the Diagnostic Molecular Pathology Laboratory of the University of California at Los Angeles sequenced select exons, and the Clinical Molecular Genetics Laboratory at Fox Chase Cancer Center (Philadelphia) did some mutation scanning. Nine laboratories (including those mentioned above) offered limited testing of the three mutations common among the Ashkenazi Jewish

population. Only Myriad, however, conducted comprehensive analysis of both BRCA genes, the technology which was most likely to be used. Source: GeneTests, BRCA1 Hereditary Breast/Ovarian Cancer: Select all clinical laboratories, at http://www.genetests.org.

13. Callon 1986.

14. Myriad Genetics, Inc., "Myriad Genetics and LabCorp Form Exclusive Predictive Medicine Marketing Alliance," press release, 2001.

15. Myriad Genetics, Inc., "Genetic Susceptibility to Breast and Ovarian Cancer: A Professional Education Program," 1999.

16. American Medical Association, "The Role of Genetic Susceptibility Testing for Breast & Ovarian Cancer," April 1999.

17. Greider 2003; Barlett and Steele 2004.

18. Official at USC Cancer Genetics Clinic, interview by the author, November 1999.

19. US Geneticist No. 6, interview by the author, November 1999.

20. Rettig 2000.

21. Lobby poster, USC-Mount Zion Cancer Center, November 1999.

22. Myriad Genetic Counselor No. 1, interview by the author, April 3, 2000.

23. Ibid.

24. Ibid.

25. Byron Spice, "Genes may become big business," *Pittsburgh Post-Gazette*, July 16, 2000. See also Merz et al. 2002.

26. National Cancer Institute, Memorandum of Understanding, 2000.

27. Myriad Genetics, Inc., "Myriad Genetics, Inc. and National Cancer Institute Forge Breast Cancer Testing Agreement," 2000.

28. National Cancer Institute official, interview by the author, December 1999.

29. Latour 1988.

30. Barbara Brenner, "The future is now," *Breast Cancer Action Newsletter*, February-March 1997.

31. Victor Volland, "Scientists isolate gene causing type of breast cancer," *St. Louis Post-Dispatch*, September 15, 1994.

32. Representative of National Alliance of Breast Cancer Organizations, interview by the author, June 2000.

33. Myriad Genetics Laboratories, advertisement for BRACAnalysis, *New York Times Magazine*, September 1999; Myriad Genetic Laboratories, advertisement for BRAC-Analysis, *US Airways Attaché*, January 2000.

34. Myriad Genetics Inc., "Breast and Ovarian Cancer: Given the Choice, Wouldn't You Rather Deal with the Known Than the Unknown?" (2000).

35. Myriad Genetics, Inc., *Testing for Hereditary Risk of Breast & Ovarian Cancer: Is It Right for You?* (VHS tape, 1999).

36. Myriad Genetics, Inc., "Media Guide Atlanta: BRACAnalysis," 2002; "Media Guide Denver: BRACAnalysis," 2002; "BRACAnalysis: Be ready against cancer" (television advertisement), 2002.

37. Myriad Genetics, Inc., "Myriad Genetics Launches Direct to Consumer Advertising for Breast Cancer Test," press release, September 12, 2002.

38. Although Myriad emphasized the woman's right to choice, this choice was severely constrained by economics. Myriad's testing services cost anywhere from $500 to $4,000, the most common full sequence analysis of both genes costing about $3,000.

39. Myriad Genetics, Inc., "Myriad Genetics Signs Agreement with Aetna US Healthcare to Provide Cancer Testing to Its Members," press release, August 3, 1998.

40. Yahoo! Finance, "Historical Prices-MYGN." Downloaded September 16, 2005 from http://table.finance.yahoo.com.

41. Aetna US Healthcare, "Prior Authorization Request Form for Breast and Ovarian Genetic Testing," 1999.

42. Kaiser Permanente, "Clinical Practice Guidelines for Referral for Genetic Counseling for Inherited Susceptibility for Breast and Ovarian Cancer," October 1997; BlueCross BlueShield Association Technology Evaluation Center, "Genetic Testing for Inherited BRCA1 or BRCA2 Mutations," June 1997.

43. Lee et al. 2002; Chaliki et al. 1995; US Geneticist No. 3, interview by the author, October 26, 1999.

44. Doksum et al. 2003.

45. Breast Cancer Action representative, interview by the author, October 1999; National Breast Cancer Coalition, "Position Statement on Genetic Testing for Inherited Predisposition to Breast Cancer," April 2004; Myriad Genetics representative, public lecture at Northwestern University, February 24, 2004; Amy Harmon, "Couples Cull Embryos to Halt Heritage of Cancer," *New York Times*, September 3, 2006.

46. Public Health Genetics Unit representative, interview by the author, July 19 1999.

47. Mackay et al., "Familial Breast Cancer."

48. R&D Office of the Anglia and Oxford NHS Executive and Unit for Public Health Genetics, "Report of Consensus Meeting on the Management of Women with a Family History of Breast Cancer," 1998.

49. Department of Health, Wetherby, "The Patient's Charter and You—a Charter for England," 1996.

50. Mackay et al., "Familial Breast Cancer."

51. R&D Office of the Anglia and Oxford NHS Executive and Unit for Public Health Genetics, "Report of Consensus Meeting," 1998.

52. Mackay et al., "Familial Breast Cancer."

53. R&D Office of the Anglia and Oxford NHS Executive and Unit for Public Health Genetics, "Report of Consensus Meeting," 1998.

54. Ibid. Other geneticists proposed alternative risk classification models to the PHGU. The UK CFSG, with the help of Ros Eeles and Gareth Evans, proposed such a model. These alternative models, however, did not quarrel with the risk assessment and triage strategy, only the substance of these categories.

55. R&D Office of the Anglia and Oxford NHS Executive and Unit for Public Health Genetics, "Report of Consensus Meeting," 1998.

56. Ibid.

57. Public Health Genetics Unit representative, interview by the author, July 19, 1999.

58. Vogel 1986.

59. Genetic Interest Group representative, interview by the author, July 6, 1998. GIG also noted that they approved of the way testing was provided through the national system: "Now our view on testing is, everything is subject to the informed consent of participants, if somebody has reason to suspect that they may be at risk from a genetic disorder, then they should have access to services . . . That testing should be done in a context where information is provided, where counseling is provided before, and after the testing process."

60. Genetic Interest Group representative, interview by the author, July 6, 1998.

61. Lois Rogers, "Breast test may save daughter from surgery," *Sunday Times*, (London), September 25, 1997; Jeremy Laurance, "Women at risk of cancer agonise over mastectomy," *Times* (London), February 26, 1996; Roger Dobson, "The body in question is mine," *Independent* (London), July 10, 1997; Rachel Baird, "Insurers act on gene test blight," *Express on Sunday* (London), February 11, 2001; Nick Thorpe, "Where should the line be drawn on who owns life itself?" *Scotsman*, July 23, 1997.

62. Wendy Watson, interview by the author, 1998.

63. Ibid.

64. Ibid.

65. NHS National Institute for Clinical Excellence, "Clinical Guideline 14: Familial Breast Cancer: The Classification and Care of Women at Risk of Familial Breast Cancer in Primary, Secondary, and Tertiary Care," 2004.

66. Ibid.

Chapter 4

1. Fosket 2004.

2. Myriad Genetics, "I did something today to guard against cancer . . . ," advertisement, *New York Times Magazine*, 1999.

3. Myriad Genetic Laboratories, "No Mutation Detected," BRACAnalysis Comprehensive BRCA-BRCA2 Gene Sequence Analysis Result, 2004.

4. Lock 1998.

5. Myriad Genetic Laboratories, "BRACAnalysis Technical Specifications," October 4, 2001.

6. Myriad Genetic Laboratories, "Testing for Hereditary Cancer Risk: What Does a Negative Result Mean?" (2000).

7. Myriad Genetic Laboratories, "Positive for a Deleterious Mutation," BRAC-Analysis Comprehensive BRCA-BRCA2 Gene Sequence Analysis Result, 2003.

8. Neuhausen et al. 1996; Easton et al. 1995; Fodor et al. 1998.

9. Myriad Genetic Laboratories, "Genetic Susceptibility to Breast and Ovarian Cancer: A Professional Education Program."

10. Ibid.

11. American Medical Association, "The Role of Genetic Susceptibility Testing for Breast & Ovarian Cancer," 1999.

12. Myriad Genetics, "Does breast or ovarian cancer run in your family?" Advertisement, *Colorado Home & Life*, 2003.

13. Myriad Genetic Laboratories, "Genetic Variant of Uncertain Significance," BRACAnalysis Comprehensive BRCA-BRCA2 Gene Sequence Analysis Result, 2001.

14. American Medical Association, "Identifying and Managing Hereditary Risk for Breast and Ovarian Cancer," 2001.

15. Myriad Genetics, advertisement for BRCA Testing, *American Journal of Human Genetics*, 1996.

16. Myriad Genetic Laboratories, "A Clinical Resource for Health Care Professionals," 1998.

17. Lorde 1980; Leopold 1999; Edare K. Carroll, "Amputation does not a cure make," *Breast Cancer Action Newsletter* 53 (April-May 1999); National Breast Cancer Coalition Fund, "NBCCF's Analysis of Health Management Options for Women With BRCA Mutations," 2005.

18. Myriad Genetics, "A Clinical Resource for Health Care Professionals," 1998.

19. Fisher et al. 1998.

20. Gail et al. 1989.

21. Dave Curtin, "Landmark study hits home here," *Denver Post*, April 7, 1988; Susan Okie, "Tamoxifen lowers risk of breast cancer," *Washington Post*, April 7, 1998; "Breast cancer breakthrough," *New York Times*, April 8, 1998.

22. "Tamoxifen: Breakthrough breast cancer drug gives hope to thousands," *Houston Chronicle*, April 10, 1998.

23. Glenda Cooper, "Drug can halve the risk of breast cancer," *Independent* (London), April 7, 1998.

24. Maura Lerner, "Breast cancer drug brings promise and tough decisions," *Minneapolis Star Tribune*, April 12, 1998.

25. Gina Kolata and Lawrence M. Fisher, "Drugs to fight breast cancer near approval," *New York Times*, September 3, 1998.

26. National Breast Cancer Coalition, "National Breast Cancer Coalition Comment on FDA Approval of Tamoxifen," press release, 1998; Robert Pear, "Preventive use of tamoxifen is allowed," *New York Times*, October 30, 1998.

27. Epstein 1996.

28. Clarke et al. 2003.

29. Fisher et al. 1998.

30. Myriad Genetics Inc., "Genetic Susceptibility to Breast and Ovarian Cancer: A Professional Education Program," 1999.

31. US Geneticist No. 4, interview by the author, August 2000

32. Myriad Genetics, Inc., "Myriad Genetics Introduces New Breast Cancer Testing Service," 1999.

33. Myriad Genetics, Inc., "The Only Thing Worse Than Hearing You Have Cancer, Is Hearing It Twice," 2000.

34. Breast Cancer Action, "BCA's policy on genetic testing for breast cancer susceptibility," *Breast Cancer Action Newsletter*, June 1996.

35. "Report of Consensus Meeting on the Management of Women with a Family History of Breast Cancer" (London and Cambridge: R&D Office of the Anglia and Oxford NHS Executive and Unit for Public Health Genetics, 1998).

36. Ibid.

37. National Institute for Clinical Excellence 2004.

38. Ibid.

39. R&D Office of the Anglia and Oxford NHS Executive and Unit for Public Health Genetics, "Report of Consensus Meeting on the Management of Women with a Family History of Breast Cancer."

40. Sarah Boseley, "US breast cancer claims denounced," *Guardian* (London), April 8, 1998.

41. Ibid.

42. Ian Murray, "Doctors accuse US of wrecking cancer drug test," *Times* (London), April 8, 1998.

43. Ibid.

44. Boseley, "US breast cancer claims denounced."

45. Powles et al. 1998; Veronesi et al. 1998.

46. Powles et al. 1998; Veronesi et al. 1998.

47. Judy Foreman, "Studies question tamoxifen data," *Boston Globe*, July 11, 1998.

48. Ibid.; Ridgely Ochs, "Studies question tamoxifen as preventative drug," *Seattle Times*, July 10, 1998.

49. R&D Office of the Anglia & Oxford NHS Executive and Unit for Public Health Genetics, "Report of Consensus Meeting on the Management of Women with a Family History of Breast Cancer."

50. "Tamoxifen Reported to Protect against Contralateral Breast Cancer in BRCA1 and BRCA2 Mutation Carriers." Downloaded July 28, 2001 from http://www.medinfo.cam.ac.uk.

51. UK Oncologist No. 2, interview by the author, August 5, 1999.

52. NICE Guidelines.

53. Julia Finch, "New market for cancer drug," *Guardian* (London), August 7, 1998.

54. Victoria Griffith, "Zeneca drug gets boost from US," *Financial Times*, September 3, 1998.

55. King et al. 2001.

56. Mackay et al., "Familial Breast Cancer."

57. Ibid.

58. UK Oncologist No. 2, interview by the author, August 5, 1999.

59. Miller et al. 2002; Humphrey et al. 2002.

60. Jennifer Trueland, "Doctors say Americans jeopardizing cancer work," *Scotsman*, April 8, 1998.

61. Doksum et al. 2003.

62. UK Oncologist No. 1, interview by the author, September 20, 1999.

Chapter 5

1. Myriad Genetics, Inc., "Myriad Genetics Launches Molecular Diagnostic Testing in Canada," press release, March 9, 2000; Eggertson 2002; Myriad Genetics, Inc., "Myriad Genetics Launches Predictive Medicine Testing in Germany, Switzerland, and Austria," press release, June 27, 2001; Myriad Genetics, Inc., "Myriad Genetics Launches Genetic Testing in Japan," press release, February 1, 2000.

2. These patents were later granted, although they have since been challenged at the European Patent Office. I discuss this challenge and its resolution in the epilogue. See "Method for diagnosing a predisposition to breast or ovarian cancer," European Patent 0699754, issued 1996; "Chromosome 13-linked breast cancer susceptibility gene," European Patent 1260520, issued 2002.

3. Myriad Genetics, "Myriad Genetics Hosting Conference of European Experts on Breast Cancer Genetic Testing," press release, 1998.

4. UK Geneticist No. 1, interview by the author, August 10, 1999.

5. Myriad Genetics, "Myriad Genetics Hosting Conference of European Experts on Breast Cancer Genetic Testing," press release, 1998.

6. UK Genetics Nurse, interview by the author, September 1999. British health-care professionals and even government officials suspected that Myriad wanted to shut down the NHS's testing services before those of any other countries because Britain likely had the most advanced testing services as well as high incidences of breast cancer. Such tactics had been used successfully in the United States in the case of Canavan's Disease (Merz 2002).

7. Department of Health official, interview by the author, October 1999.

8. Nick Thorpe, "Where should the line be drawn on who owns life itself?" *Scotsman*, July 23, 1997; Roger Dobson, "Women fight patent on cancer test," *Sunday Times*, April 20, 1997; Roger Dobson, "The body in question is mine," *Independent* (London), July 10, 1997.

9. Department of Health official, interview by the author, October 1999.

10. Parthasarathy 2003. Jeremy Rifkin, an American critic of biotechnology, did start a petition opposing the patenting of the BRCA genes. (See Tamar Lewin, "Move to patent gene is called obstacle to research," *New York Times*, May 21, 1996.) This petition was signed by some patient activist groups, including the National Breast Cancer Coalition and the National Ovarian Cancer Coalition, but it did not result in any major policy change. In fact, the NBCC later printed an article in its newsletter announcing an equivocal position on the patenting of human genes. It argued that patenting could be beneficial to research, whereas a ban on patenting genes might have unforeseen consequences ("Gene patenting: Yes or no?" *Call to Action*, fall-winter 1997). While American scientists and physicians voiced little opposition to gene patenting in the 1980s and the early 1990s, some groups began to criticize the practice when researchers attempted to patent expressed sequence tags (gene fragments with unknown function). These criticisms, however, did not usually cover the patenting of genes with known function—e.g., disease genes (American Medical Association, "Patenting of genes and their mutations," Report 9, Council on Scientific Affairs, January 2000). Controversy over the patenting of genes has continued to grow, and the December 2002 issue of *Academic Medicine*, the journal of the American Association of Medical Colleges, was devoted entirely to this topic. It should be noted, however, that opposition to gene patenting has been much more straightforward and vigorous in Britain than in the United States. (See "Special theme: Public versus private ownership of scientific discovery," *Academic Medicine* 77, December 2002: 1179–1399.)

11. American Society of Human Genetics, "Position Paper on Patenting of Expressed Sequence Tags," November 1991; American Society of Human Genetics to Assistant Commissioner for Patents, US Patent and Trademark Office, "RE: Revised interim utilities examination guidelines and revised interim guidelines for examination of patent application," letter, March 22, 2000.

12. Kleinman 2003.

13. Magnus et al. 2002.

14. U.S. Congress, House of Representatives, Committee on the Judiciary, Subcommittee on Courts, the Internet, and Intellectual Property, hearing on gene patents and other genomic inventions, July 13, 2000; Dr. Edward R. B. McCabe, Chair, Secretary's Advisory Committee on Genetic Testing, to Dr. Donna Shalala, U.S.

Secretary of Health and Human Services, November 17, 2000: "On behalf of the Secretary's Advisory Committee on Genetic Testing (SACGT), I am writing to bring your attention an issue related to access to genetic tests. . . ."

15. In 1980, the US Congress passed the Bayh-Dole Act (Public Law 96–517), which allowed for private ownership of inventions funded by the taxpayer. Small businesses, non-profit organizations, and universities could now patent and commercialize inventions that resulted from government-funded research. Many universities quickly created specially designated technology-transfer offices, and today many reap significant revenues from these inventions. See Kenney 1986.

16. *Moore v. Regents of the University of California*, 793 P2d 479, 271 Cal. Rptr. 146 (1990).

17. Hilgartner 2004.

18. de Laet 2000.

19. von Gabain and Lanthaler 2003.

20. European Patent Office, "'Oncomouse' Opposition Proceedings Resume at EPO," press release, November 5, 2001; Paula Park, "EPO Restricts OncoMouse Patent," The Scientist: Daily News (email service), July 26, 2004.

21. Ibid.

22. Nigel Hawkes, "Euro MPs turn down life-form patent law," *Times* (London), March 2, 1995; Daniel Green, "Parliament scuppers a new patents directive," *Financial Times*, March 3, 1995; Charles Bremmer, "Euro-MPs clear way for genetic patents," *Times* (London), May 13, 1998; Katherine Butler and Charles Arthur, "Anger as Europe votes to 'sell off' genes," *Independent* (London), May 13, 1998.

23. British Society for Human Genetics, "Patenting of Human Gene Sequences and the EU Draft Directive," September 1997. Downloaded April 1, 2001 from http://www.bshg.org.uk.

24. Tony Andrews et al., letter to European Parliament, July 14, 1997.

25. British Society for Human Genetics, "Patenting of Human Gene Sequences." Downloaded April 1, 2001 from http://www.bshg.org.uk.

26. Wendy Watson, interview by the author, June 1998.

27. Church of Scotland, "Church of Scotland Urges European Parliament to Prevent Patenting Human Genes," press release, May 22, 1997.

28. European Parliament and Council of the European Union, "Directive on the Legal Protection of Biotechnological Inventions," Directive 98/44/EC, 1998.

29. Dickson 2000.

30. British Society for Human Genetics, "BSHG Statement on Patenting and Clinical Genetics." Downloaded December 8, 1999 from http://www.bham.ac.uk.

31. Ibid.

32. Clinical Molecular Genetics Society, "Gene Patents and Clinical Molecular Genetic Testing in the UK," p. 1.

33. James Meek, "Money and the meaning of life," *Guardian* (London), January 17, 2000.

34. Emma Ross, "Scientists object to gene patent," Associated Press, January 18, 2000.

35. Ibid.

36. The sociologist of science Robert Merton argued that disinterestedness and communitarianism were among the norms common to the scientific profession. Michael Mulkay responded to these "Mertonian norms," arguing that they reflected the profession's ideal image of itself rather than governing the behavior of day-to-day scientists. Ian Mitroff took this argument further, arguing that scientists were indeed personally and professionally invested in their work and could be described by a set of "counter-norms" that included emotional commitment and interestedness. See Merton 1973; Mulkay 1976; Mitroff 1974.

37. Steve Connor, "Concern over cancer gene patent," *Independent* (London), September 15, 1994.

38. Wendy Watson, interview by the author, October 1999.

39. Davies and White 1996.

40. Dalpé et al. 2003.

41. UK Geneticist No. 1, interview by the author, August 10, 1999.

42. British Society for Human Genetics, "Patenting of Human Gene Sequences and the EU Draft Directive," 1997.

43. UK Geneticist No. 2, interview by the author, July 27, 1998.

44. UK Geneticist No. 3, interview by the author, August 13, 1998.

45. Roger Dobson, "Women fight patent on cancer test," *Sunday Times* (London), April 20, 1997.

46. Wendy Watson, interview by the author, June 1998.

47. UK Molecular Geneticist, interview by the author, December 23, 1999.

48. Clinical Molecular Genetics Society, "Gene Patents and Clinical Molecular Genetic Testing in the UK," 1999.

49. The idea that accuracy is a social achievement that is locally contingent and requires negotiation over standards and margins of error has been widely discussed by scholars of science and technology studies. See Pinch 1993.

50. James Meek, "Money and the meaning of life," *Guardian* (London), January 17, 2000.

51. UK Oncologist No. 1, interview by the author, September 20, 1999.

52. MacKenzie 1993.

53. UK molecular geneticist, interview by the author, December 23, 1999.

54. Clinical Molecular Genetics Society, "Gene Patents and Clinical Molecular Genetic Testing in the UK," 1999.

55. Ibid.

56. Ibid.

57. Ibid.

58. Wendy Watson, interview by the author, October 1999.

59. Steve Connor, "Concern over cancer gene patent," *Independent* (London). September 15, 1994.

60. Watson, interview by the author.

61. Sheila Adam, "Future Provision of BRCA-Testing Services," letter, London, 2000.

62. Myriad Genetics, "Myriad Genetics Launches Genetic Testing in the United Kingdom and Ireland," press release, March 2000.

63. Rosgen Ltd., "UK Company Announces Licensing Agreement for Breast Cancer Genetic Testing," press release, March 2000.

64. Southwest Thames Regional Genetics Service, Stop Press: Myriad Genetics Attempt to Monopolize Breast Cancer Testing." Downloaded September 21, 2005 from http://www.genetics-swt.org.

65. Rosgen Ltd., "UK Company Announces Licensing Agreement for Breast Cancer Genetic Testing," press release, March 2000.

Conclusion

1. Walsh et al. 2006.

2. DNA Direct, "Hereditary Breast & Ovarian Cancer: How Testing Works." Downloaded September 21, 2005 from http://www.dnadirect.com.

3. LabCorp, "Hereditary Cancer Syndromes: The Power of Predictive Medicine," 2002.

4. DNA Direct, "Hemochromatosis: How Testing Works." Downloaded September 21, 2005 from http://www.dnadirect.com.

5. US Preventive Services Task Force 2005.

6. MediChecks.com, "BRCA1/BRCA2." Downloaded September 21, 2005 from http://www.medichecks.com.

7. Human Genetics Commission, "Genes Direct: Ensuring the Effective Oversight of Genetic Tests Supplied Directly to the Public" (London: Department of Health, 2003).

8. Nuffield Council on Bioethics, "Pharmacogenetics: Ethical Issues, 2003."

9. Sciona, Inc. "How Cellf Works." Downloaded April 5, 2006 from http://www.mycellf.com.

10. Duster 2003; King 1999; Robertson 2003; Dahl 2003; de Wert 2005.

11. Walsh et al. 2006.

12. Schot and Rip 1996. See also Guston and Sarewitz 2002.

Epilogue

1. UK Geneticist No. 5, interview by the author, August 11, 1998.

2. European Parliament, "European Parliament Resolution on the Patenting of BRCA1 and BRCA2 ('Breast Cancer') Genes, B5–0633, 0641, 0651 and 0663/2001," 2001.

3. Institut Curie, "The Institut Curie, the Assistance Publique-Hôpitaux de Paris and the Institut Gustave-Roussy File a Joint Opposition Notice to the Myriad Genetics Patent with the European Patent Office," press release, October 10, 2001.

4. Matthijs and Halley 2002.

5. European Parliament, "European Parliament Resolution on the Patenting of BRCA1 and BRCA2 ('Breast Cancer') Genes, B5–0633, 0641, 0651 and 0663/2001."

6. Benowitz 2003.

7. Gad et al. 2001.

8. Benowitz 2003.

9. Gad et al. 2001.

10. Matthijs and Halley 2002.

11. Greenpeace, "Streit um Patente auf Brustkrebs-Gene," press release, January 2005.

12. Institut Curie, "Against Myriad Genetic's Monopoly on Tests for Predisposition to Breast and Ovarian Cancer Associated with the BRCA1 gene," press release, September 26, 2002.

13. Grit Kienzlen, "BRCA2 Patent Upheld." Downloaded July 15, 2005 from http://www.the-scientist.com.

14. Bioscientia, "Genetic Testing for Hereditary Breast and Ovarian Cancer: BRCA1/2 Analysis." Downloaded September 16, 2005 from http://www.bioscientia.de.

Bibliography

Abbott, Andrew. 1988. *The System of the Professions: An Essay on the Division of Expert Labor*. University of Chicago Press.

Akrich, Madeline. 1992. "The de-scription of technical objects." In *Shaping Technology/Building Society*, ed. W. Bijker and J. Law. MIT Press.

American Society of Human Genetics. 1994. "Statement of the American Society of Human Genetics on genetic testing for breast and ovarian cancer predisposition." *American Journal of Genetics* 55: i–iv.

Anderson, Christopher. 1991. "US patent application stirs up gene hunters." *Nature* 353, October: 485–486.

Andrews, Lori, et al. 1994. *Assessing Genetic Risks: Implications for Health and Social Policy*. National Academy Press.

Annan, Kofi. 2003. "A challenge to the world's scientists." *Science* 299: 1485.

Árnason, Arnar, and Bob Simpson. 2003. "Refractions through culture: The new genomics in Iceland." *Ethnos* 68: 533–553.

Ashmore, Malcom, Michael Mulkay, and Trevor Pinch. 1989. *Health and Efficiency: A Sociology of Health Economics*. McGraw-Hill.

Barlett, Donald, and James Steele. 2004. *Critical Condition: How Health Care in America Became Big Business—and Bad Medicine*. Doubleday.

Bauer, M., ed. 1995. *Resistance to New Technology*. Cambridge University Press.

Benichou, Jacques, Mitchell Gail, and John Mulvihill. 2004. "Graphs to estimate an individualized risk of breast cancer." *Journal of Clinical Oncology* 14: 103–110.

Benowitz, Steve. 2003. "European groups oppose Myriad's latest patent on BRCA1." *Journal of the National Cancer Institute* 95: 8–9.

Bijker, Wiebe. 1997. *Of Bicycles, Bakelite, and Bulbs: Toward a Theory of Sociotechnical Change*. MIT Press.

Bijker, Wiebe, Thomas Hughes, and Trevor Pinch, eds. 1987. "General introduction." In *The Social Construction of Technological Systems: New Directions in the Sociology and History of Technology*. MIT Press.

Bijker, Wiebe, Thomas Hughes, and Trevor Pinch. 1989. *The Social Construction of Technological Systems*. MIT Press.

Boston Women's Health Book Collective. 1973. *Our Bodies, Ourselves*. Simon and Schuster.

Brickman, Ronald. 1982. *Chemical Regulation and Cancer: A Cross-National Study of Policy and Politics*. National Technical Information Service, US Department of Commerce.

Brickman, Ronald, Sheila Jasanoff, and Thomas Ilgen. 1982. *Chemical Regulation and Cancer: A Cross-National Study of Policy and Politics*. Cornell University Press

Brickman, Ronald, Sheila Jasanoff, and Thomas Ilgen. 1985. *Controlling Chemicals: The Politics of Regulation in Europe and the United States*. Cornell University Press.

Burke, Wylie, Mary Jo Ellis Kahn, Judy Garber, and Francis Collins. 1996. "'First do no harm' also applies to cancer susceptibility testing." *Cancer Journal* 2, no. 5: 250–252.

Callon, Michel.1986. "Elements of a sociology of translation: Domestication of the Scallops and the Fishermen of St Brieuc Bay." In *Power, Action and Belief*, ed. J. Law. Routledge.

Callon, Michel. 1987. "Society in the making: The study of technology as a tool for sociological analysis." In *The Social Construction of Technological Systems*, ed. W. Bijker et al. MIT Press.

Casamayou, Maureen. 2001. *The Politics of Breast Cancer*. Georgetown University Press.

Casper, Monica J., and Adele E. Clarke. 1995. "Making the Pap smear into the 'right tool' for the job: Cervical cancer screening in the United States, circa 1940–95." *Social Studies of Science* 28: 255–290.

Chaliki, Hemasree, Starlene Loader, Jeffrey Levenkron, Wende Logan-Young, W. Jackson Hall, and Peter Rowley. 1995. "Women's receptivity to testing for a genetic susceptibility to breast cancer." *American Journal of Public Health* 85: 1133–1135.

Chang, Virginia, and Nicholas Christakis. 2002. "Medical modeling of obesity: A transition from action to experience in a twentieth century American textbook." *Sociology of Health and Illness* 24: 151–177.

Cho, Mildred, Samantha Illangasekare, Meredith Weaver, Debra Leonard, and Jon Merz. 2003. "Effects of patents and licenses on the provision of clinical genetic testing services." *Journal of Molecular Diagnostics* 5: 3–8.

Clarke, Adele, Janet Shim, Larua Mamo, Jennifer Ruth Fosket, and Jennifer Fishman. 2003. "Biomedicalization: Technoscientific transformations of health, illness, and US biomedicine." *American Sociological Review* 68, April: 161–194.

Claus, Elizabeth, Neil Risch, and W. Douglas Thompson. 1994. "Autosomal dominant inheritance of early-onset breast cancer, implications for risk prediction." *Cancer* 73: 643–651.

Coventry, Peter, and John Pickstone. 1999. "From what and why did genetics emerge as a medical specialism in the 1970s in the UK?" *Social Science & Medicine* 49: 1227–1238.

Dahl, Edgar. 2003. "Ethical issues in new uses of preimplantation genetic diagnosis." *Human Reproduction* 18: 1368–1369.

Dalpé, Robert, Louise Bouchard, Anne-Julie Houle, and Louis Bédard. 2003. "Watching the race to find the breast cancer genes." *Science, Technology, and Human Values* 28: 187–216.

Davies, Kevin, and Michael White. 1996. *Breakthrough: The Race to Find the Breast Cancer Gene.* Wiley.

de Laet, Marianne. 2000. "Patents, travel, space: Ethnographic encounters with objects in transit." *Environment and Planning: Society and Space* 18, no. 2: 152.

de Wert, Guido. 2005. "Preimplantation genetic diagnosis: The ethics of intermediate cases." *Human Radiation* 20: 3261–3266.

Dickson, David. 2000. "Politicians seek to block human-gene patents in Europe." *Nature* 404, no. 6780: 802.

Doksum, Teresa, Barbara Bernhardt, and Neil Holtzman. 2003. "Does knowledge about the genetics of breast cancer differ between nongeneticist physicians who do or do not discuss or order BRCA testing?" *Genetics in Medicine* 5: 99–105.

Duster, Troy. 1990. *Backdoor to Eugenics.* Routledge.

Duster, Troy. 2003. *Backdoor to Eugenics,* second edition. Routledge.

Easton, D. 1994. "Cancer risks in A-T heterozygotes." *International Journal of Radiation Biology* 66: 177–182.

Easton, D., D. Ford, and D. Bishop. 1995. "Breast and ovarian cancer incidence in BRCA1-mutation carriers." *American Journal of Human Genetics* 56: 265–271.

Edquist, Charles. 1997. *Systems of Innovation: Technologies, Institutions, and Organizations.* Pinter.

Eggertson, Laura. 2002. "Ontario defies US firm's genetic patent, continues cancer screening." *Canadian Medical Association Journal* 166: 494.

Eisenberg, Rebecca S. 2002. "How Can You Patent Genes?" *American Journal of Bioethics* 2, no. 3: 26–28.

Epstein, Steven. 1996. *Impure Science: AIDS, Activism, and the Politics of Knowledge.* University of California Press.

Fisher, Bernard, et al. 1998. "Tamoxifen for prevention of breast cancer: Report of the National Surgical Adjuvant Breast and Bowel Project P-1 Study." *Journal of the National Cancer Institute* 90: 1371–1388.

Fodor, Flora, Ainsley Weston, Ira Bleiweiss, Leslie McCurdy, Mary Walsh, Paul Tartter, Steven Brower, and Christine Eng. 1998. "Frequency and carrier risk associated with common BRCA1 and BRCA2 mutations in Ashkenazi Jewish breast cancer patients." *American Journal of Human Genetics* 63: 45–51.

Ford, D., et al. 1998. "Genetic heretogeneity and penetrance analysis of the BRCA1 and BRCA2 genes in breast cancer families." *American Journal of Human Genetics* 62: 676–689.

Fosket, Jennifer. 2004. "Constructing 'high-risk women': The development and standardization of a breast cancer risk assessment tool." *Science, Technology, and Human Values* 29: 291–313.

Fox, Renée. 1977. "The medicalization and demedicalization of American society." *Daedalus* 106: 9–22.

Friedan, Betty. 1993. *The Fountain of Age.* Touchstone.

Friedan, Betty. 2001. *The Feminine Mystique.* Norton.

Gad, Sophie, Maren Scheuner, Sabine Pages-Berhouet, Virginie Caux-Moncoutier, Aaron Bensimon, Alain Aurias, Mark Pinto, and Dominique Stoppa-Lyonnet. 2001. "Identification of a large rearrangement of the BRCA1 gene using colour bar code on combed DNA in an American breast/ovarian cancer family previously studied by direct sequencing." *Journal of Medical Genetics* 38: 388–392.

Gail, Mitchell, Louise Brinton, David Byar, Donald Corle, Sylvan Green, Catherine Schairer, and John Mulvihill. 1989. "Projecting individualized probabilities of developing breast cancer for white females who are being examined annually." *Journal of the National Cancer Institute* 81: 1879–1886.

Ganguly, Arupa, Matthew Rock, and Darwin Prockop. 1993. "Conformation sensitive gel electrophoresis for rapid detection of single base differences in double-stranded PCR products and DNA fragments: Evidence for solvent induced bends in DNA heteroduplexes." *Proceedings of the National Academy of Sciences* 90: 10325–10329.

Gieryn, Thomas. 1983. "Boundary-work and the demarcation of science from non-science: Strains and interests in professional ideologies of scientists." *American Sociological Review* 48: 781–795.

Gottweis, Herbert. 2002. "Stem cell policies in the United States and in Germany: Between bioethics and regulation." *Policy Studies Journal* 30: 444–470.

Greider, Katherine. 2003. *The Big Fix: How the Pharmaceutical Industry Rips Off American Consumers*. Public Affairs.

Guston, David, and Daniel Sarewitz. 2002. "Real-time technology assessment." *Technology in Society* 24: 93–109.

Hall, Jeff, Ming Lee, Beth Newman, Jan Morrow, Lee Anderson, Bing Huey, and Mary-Claire King. 1990. "Linkage of early-onset familial breast cancer to chromosome 17q21." *Science* 250: 1684–1689.

Hedgecoe, Adam. 1998. "Geneticization, medicalisation, and polemics." *Medicine, Healthcare, and Philosophy* 4, no. 3: 305–309.

Hilgartner, Stephen. 2000. *Science on Stage: Expert Advice as Public Drama*. Stanford University Press.

Hilgartner, Stephen. 2004. "The constitution of genomic property: Co-producing mapping strategies and moral orders in genome laboratories." In *States of Knowledge*, ed. S. Jasanoff. Harvard University Press.

Holtzman, Neil, and David Shapiro. 1998. "Genetic testing and public policy." *British Medical Journal* 316: 852–856.

Hsia, Y., K. Hirshhorn, R. Silverberg, and L. Godmilow, eds. 1979. *Counseling in Genetics*. Liss.

Hubbard, Ruth, and Richard Lewontin. 1996. "Sounding board: Pitfalls of genetic testing." *New England Journal of Medicine* 334: 1192–1193.

Hughes, Thomas. 1983. *Networks of Power: Electrification in Western Society, 1880–1930*. Johns Hopkins University Press.

Humphrey, Linda, Mark Helfand, Benjamin Chan, and Steven Woolf. 2002. "Breast cancer screening: A summary of the evidence for the US preventive task force." *Annals of Internal Medicine* 137, no. 5: 305–312.

Institute of Medicine, Division of Health Sciences Policy, Committee on Assessing Genetic Risks. 1994. *Assessing Genetic Risks: Implications for Health and Social Policy*. National Academy Press.

Jasanoff, Sheila. 1991. "Acceptable evidence in a pluralistic society." In *Acceptable Evidence*, ed. R. Hollander and D. Mayo. Oxford University Press.

Jasanoff, Sheila. 1995. "Product, process, or programme: Three cultures and the regulation of biotechnology." In *Resistance to New Technology*, ed. M. Bauer. Cambridge University Press.

Jasanoff, Sheila. 2005. *Designs on Nature*. Princeton University Press.

Kenney, Martin. 1986. *Biotechnology: The University-Industrial Complex*. Yale University Press.

Kenney, Martin. 1998. "Biotechnology and the creation of a new economic space." In *Private Science*, ed. A. Thackray. University of Pennsylvania Press.

Kevles, Daniel. 1998. *In the Name of Eugenics: Genetics and the Uses of Human Heredity* (reprint edition). Harvard University Press.

King, Mary-Claire, Sam Wieand, Kathryn Hale, Ming Lee, Tom Walsh, Kelly Owens Joanathan Tait, Leslie Ford, Barbara K. Dunn, Joseph Costantino, Lawrence Wiskerham, Norman Wolmark, and Bernard Fisher. 2001. "Tamoxifen and breast cancer incidence among women with inherited mutations in BRCA1 and BRCA2." *Journal of the American Medical Association* 286: 2251–2256.

King, David S. 1999. "Preimplantation genetic diagnosis and the 'new' eugenics." *Journal of Medical Ethics* 25: 176–182.

Klein, Rudolf. 2001. *The New Politics of the National Health Service*. Prentice-Hall.

Kleinman, Daniel. 2003. *Impure Cultures: University Biology and the World of Commerce*. University of Wisconsin Press.

Kline, Ronald, and Trevor Pinch. 1996. "Users as agents of technological change: The social construction of the automobile in the rural United States." *Technology and Culture* 37: 763–795.

Kushner, Rose. 1975, *Breast Cancer: A Personal History and an Investigative Report*. Harcourt Brace Jovanovich.

Kushner, Rose. 1980. *If You've Thought about Breast Cancer—*. Women's Breast Cancer Advisory Center.

Kushner, Rose. 1982. *Why Me?* Holt, Rinehart and Winston.

Kushner, Rose. 1985. *Alternatives: New Developments in the War on Breast Cancer*. Warner Books.

Latour, Bruno. 1988. *Science in Action: How to Follow Scientists and Engineers through Society*. Harvard University Press.

Lee, Soo-Chin, Barbara Bernhardt, and Kathy Helzlsouer. 2002. "Utilization of BRCA1/2 genetic testing in the clinical setting." *Cancer* 94: 1876–1885.

Leeson, Joyce. 1979. *Women and Medicine*. Routledge.

Leopold, Ellen. 1999. *A Darker Ribbon: Breast Cancer, Women, and Their Doctors in the Twentieth Century*. Beacon.

Lerner, Barron. 1999. "Great expectations: Historical perspectives on genetic breast cancer testing." *American Journal of Public Health* 89: 938–944.

Lerner, Barron. 2001. *The Breast Cancer Wars: Hope, Fear, and the Pursuit of a Cure in Twentieth-Century America*. Oxford University Press.

Lindee, M. Susan. 2000. "Genetic disease since 1945." *Nature Reviews Genetics* 1: 236–241.

Lippman, Abby. 1991. "Prenatal genetic testing and screening: Constructing needs and reinforcing inequities." *American Journal of Law and Medicine* 17: 15–50.

Lock, Margaret. 1998. "Breast cancer: Reading the omens." *Anthropology Today* 14: 7–16.

Lorde, Audre. 1980. *The Cancer Journals*. Aunt Lute Books.

Love, Susan, and Karen Lindsey. 2000. *Dr. Susan Love's Breast Book*, third edition. HarperCollins.

Löwy, Ilana. 1997. *Between Bench and Bedside: Science, Healing, and Interleukin-2 in a Cancer Ward*. Harvard University Press.

MacKenzie, Donald. 1993. *Inventing Accuracy: A Historical Sociology of Nuclear Missile Guidance*. MIT Press.

MacKenzie, Donald, and Judy Wajcman, eds. 1985. *The Social Shaping of Technology*. Open University Press.

Magnus, David, Arther Caplan, and Glenn McGee. 2002. *Who Owns Life?* Prometheus Books.

Matthijs, Gert, and Dicky Halley. 2002. "European-wide opposition against the breast cancer gene patents." *European Journal of Human Genetics* 10: 783–784.

Merton, Robert. 1973. "The normative structure of science." In Merton, *The Sociology of Science*. University of Chicago Press.

Merz, Jon. 2002. "Discoveries: Are there limits on what may be patented?" In *Who Owns Life?* ed. D. Magnus, A. Caplan, and G. McGee. Prometheus Press.

Merz, Jon, et al. 2002. "Diagnostic testing fails the test." *Nature* 415: 577–579.

Metcalfe, Stan. 1995. "The Economic foundations of technology policy: Equilibrium and evolutionary perspectives." In *Handbook of the Economics of Innovation and Technological Change*, ed. P. Stoneman. Blackwell.

Meyer, John, Francisco Ramirez, Evan Schofer, and Gili Drori. 2003. *Science in the Modern World Polity: Institutionalization and Globalization*. Stanford University Press.

Miki, Yoshio, Jeff Swensen, Donna Shattuck-Eidens, P. Andrew Futreal, Keith Harshman, Sean Tavtigian, Qingyun Liu, Charles Cochran, L. Michelle Bennett, Wei Ding, Russell Bell, Judith Rosenthal, Charles Hussey, Thanh Tran, Melody McClure, Cheryl Frye, Tom Hattier, Robert Phelps, Astrid Haugen-Strano, Harold Katcher,

Kazuko Yakumo, Zahra Gholami, Daniel Shaffer, Steven Stone, Steven Bayer, Christian Wray, Robert Bogden, Priya Dayananth, John Ward, Patricia Tonin, Steven Narod, Pam Bristow, Frank Norris, Leah Helvering, Paul Morrison, Paul Rosteck, Mei Lai, J. Carl Barrett, Cathryn Lewis, Susan Neuhausen, Lisa Cannon-Albright, David Goldgar, Roger Wiseman, Alexander Kamb, and Mark Skolnick. 1994. "A strong candidate for the breast and ovarian cancer susceptibility gene BRCA1." *Science* 266: 66–71.

Miller, Anthony, Teresa To, Cornelia Baines, and Claus Wall. 2002. "The Canadian National Breast Screening Study-1: Breast cancer mortality after 11 to 16 years of Follow-up." *Annals of Internal Medicine* 137, no. 5: 305–312.

Mitroff, Ian. 1974. "Norms and counter-norms in a select group of the Apollo moon scientists: A case study of the ambivalence of scientists." *American Sociological Review* 39, August: 579–595.

Mosca, Lori, Wanda Jones, Kathleen King, Pamela Ouyang, Rita Redberg, and Martha Hill. 2000. "Awareness, perception, and knowledge of heart disease risk and prevention among women in the United States." *Archives of Family Medicine* 9: 506–515.

Mulkay, Michael. 1976. "Norms and ideology in science." *Social Science Information* 15, no. 4: 637–656.

National Research Council. 2002. *Care without Coverage: Too Little, Too Late*. National Academy Press.

Nelkin, Dorothy, and M. Susan Lindee. 1996. *The DNA Mystique*. Freeman.

Nelkin, Dorothy, and Lawrence Tancredi. 1991. *Dangerous Diagnostics: The Social Power of Biological Information*. Basic Books.

Neuhausen, S., S. Mazoyer, L. Friedman, M. Stratton, K. Offit, A. Caligo, G. Tomlinson, L. Cannon-Albright, T. Bishop, D. Kelsell, E. Solomon, B. Weber, F. Couch, J. Struewing, P. Tonin, F. Durocher, S. Narod, M. Skolnick, G. Lenoir, O. Serova, B. Ponder, D. Stoppa-Lyonnet, D. Easton, M. King, and D. Goldgar. 1996. "Haplotype and phenotype analysis of six recurrent BRCA1 Mutations in 61 families: Results of an international study." *American Journal of Human Genetics* 58: 271–280.

Nyström, Lennarth, Ingvar Andersson, Nils Bjurstam, Jan Frisell, Bo Nordenskjöld, and Lars Erik Rutqvist. 2002. "Long-term effects of mammography screening: Updated overview of the Swedish randomised trials." *Lancet* 359: 909–919.

Olsen, Ole, and Peter Gøtzsche. 2001. "Cochrane review on screening for breast cancer with mammography." *Lancet* 358: 1340–1342.

Oudshoorn, Nelly, and Trevor Pinch, eds. 2003. *How Users Matter*. MIT Press.

Parthasarathy, Shobita. 2003. "Knowledge is power: Genetic testing for breast cancer and patient activism in the US and Britain." In *How Users Matter*, ed. N. Oudshoorn and T. Pinch. MIT Press.

Parthasarathy, Shobita. 2004. "Regulating risk: Defining genetic privacy in the United States and Britain." *Science, Technology, and Human Values* 29: 332–352.

Patel, Pari, and Keith Pavitt. 1995. "The nature and economic importance of national innovation systems." *STI Review* 14: 9–32.

Paul, Diane. 1995. *Controlling Human Heredity: 1865 to the Present.* Humanity Books.

Pinch, Trevor. 1993. "Testing, one, two, three—testing: Towards a sociology of testing." *Science, Technology & Human Values* 18: 25–41.

Pinn, Vivian, and Debbie Jackson. 1996. "Advisory committee to NIH office passes resolutions." *Journal of Women's Health* 5, no. 6: 549–553.

Pollock, Allyson. 2004. *NHS plc. The Privatisation of Our Health Care.* Verso.

Powles, Trevor, Ros Eeles, Sue Ashley, Doug Easton, Jenny Chang, Mitch Dowsett, Alwynne Tidy, Jenny Viggers, and Jane Davey. 1998. "Interim analysis of the incidence of breast cancer in the Royal Marsden Hospital tamoxifen randomised chemoprevention trial." *Lancet* 352: 98–101.

Rabinow, Paul. 1999. *French DNA.* University of Chicago Press.

Ragaz, Joseph, Stewart Jackson, Nhu Le, Ian Plenderleith, John Spinelli, Vivian Basco, Kenneth Wilson, Margaret Knowling, Christopher Coppin, Marilyn Paradis, Andrew Coldman, and Ivo Olivotto. 1997. "Adjuvant radiotherapy and chemotherapy in node-positive premenopausal women with breast cancer." *New England Journal of Medicine* 337, no. 14: 956–962.

Reardon, Jennifer. 2001. "The Human Genome Diversity Project: A case study in coproduction." *Social Studies of Science* 31, June: 357–389.

Rettig, Richard. 2000. "The industrialization of clinical research." *Health Affairs* 19, March-April: 129–146.

Robertson, John. 2003. "Extending preimplantation genetic diagnosis: Medical and non-medical uses." *Journal of Medical Ethics* 29: 213–216.

Rothman, David. 2003. *Strangers at the Bedside: A History of How Law and Bioethics Transformed Medical Decision Making.* Aldine.

Rothstein, Mark. 1999. *Genetic Secrets: Protecting Privacy and Confidentiality in the Genetic Era.* Yale University Press.

Ruzek, Sheryl. 1978. *The Women's Health Movement: Feminist Alternatives to Medical Control.* Praeger.

Schot, Johan, and Arie Rip. 1996. "The past and future of constructive technology assessment." *Technological Forecasting and Social Change* 54: 251.

Schulman, Joseph, and Harvey Stern. 1996. "Genetic predisposition testing for breast cancer." *Cancer Journal* 2: 244–252.

Singletary, S., and S. Kroll. 1996. "Skin-sparing mastectomy with immediate breast reconstruction." *Advances in Surgery* 30: 39–52.

Skocpol, Theda. 1997. *Boomerang: Health Care Reform and the Turn Against the Government*. Norton.

Stabiner, Karen. 1998. *To Dance with the Devil*. Delta.

Star, Susan Leigh, and James Griesemer. 1989. "Institutional ecology, 'translations' and boundary objects: Amateurs and professionals in Berkeley's Museum of Vertebrate Zoology, 1907–39." *Social Studies of Science* 19, no. 3: 393, 411.

Starr, Paul. 1984. *Social Transformation of American Medicine* (reprint edition). Basic Books.

Stratton, John, C. Hilary Buckley, David Lowe, and Bruce Ponder. 1999. "Comparison of prophylactic oophorectomy specimens from carriers and noncarriers of a BRCA1 or BRCA2 gene mutation." *Journal of the National Cancer Institute* 91, no. 7: 626–628.

Struewing, Jeffery, Patricia Hartge, Sholom Wacholder, Sonya Baker, Martha Berlin, Mary McAdams, Michelle Timmerman, Lawrence Brody, and Margaret Tucker. 1997. "The risk of cancer associated with specific mutations of BRCA1 and BRCA2 among Ashkenazi Jews." *New England Journal of Medicine* 336, no. 20: 1401–1408.

Swidler, Ann. 1986. "Culture in action: Symbols and strategies." *American Sociological Review* 51: 273–286.

Tavtigian, S., et al. 1996. "The complete BRCA2 gene and mutations in chromosome 13q-linked kindreds." *Nature Genetics* 12: 333–337.

Thorlacius, Steinunn, Jeffery Struewing, Patricia Hartge, Gudridur Olafsdottir, Helgi Sigvaldason, Laufey Tryggvadottir, Sholom Wacholder, Hrafn Tulinius, and Jorunn Eyfjörd. 1998. "Population-based study of risk of breast cancer in carriers of BRCA2 mutation." *Lancet* 352: 1337–1339.

Thomas, Sandy, Michael Hopkins, and Max Brady. 2002. "Shares in the human genome—the future of patenting DNA." *Nature Biotechnology* 20: 1185–1188.

US Preventive Services Task Force. 2005. "Genetic risk assessment and BRCA mutation testing for breast and ovarian cancer susceptibility: Recommendation statement." *Annals of Internal Medicine* 143: 355–361.

Veronesi, U., et al. 1998. "Prevention of breast cancer with tamoxifen: Preliminary findings from the Italian randomised trial among hysterectomised women." *The Lancet* 352: 93–97.

Vogel, David. 1986. *National Styles of Regulation: Environmental Policy in Great Britain and the United States*. Cornell University Press.

von Beuzekom, Brigitte. 2001. Biotechnology Statistics in OECD Member Countries: Compendium of Existing National Statistics. STI Working Paper 2001/6, Organization for Economic Cooperation and Development.

von Gabain, Alexander, and Werner Lanthaler. 2003. "European biotech hasn't hit the street." *EMBO Reports* 4, no. 7: 642–646.

Wailoo, Keith. 1997. *Drawing Blood: Technology and Disease Identity in Twentieth-Century America.* Johns Hopkins University Press.

Walsh, Tom, Silvia Casadei, Kathryn Hale Coats, Elizabeth Swisher, Sunday Stray, Jake Higgins, Kevin Roach, Jessica Mandell, Ming Lee, Sona Ciernikova, Lenka Foretova, Pavel Soucek, and Mary-Claire King. 2006. "Spectrum of mutations in BRCA1, BRCA2, CHEK2, and TP53 in families at high risk of breast cancer." *Journal of the American Medical Association* 295: 1379–1388.

Webster, Charles. 1998. *The National Health Service: A Political History.* Oxford University Press.

Wexler, Nancy. 1992. "Clairvoyance and caution: Repercussions from the Human Genome Project." In *The Code of Codes: Scientific and Social Issues in the Human Genome Project,* ed. D. Kevles and L. Hood. Harvard University Press.

Willems, Dick, Elisabeth Daniels, Gerrit van der Wal, Paul van der Maas, and Ezekiel Emanuel. 2000. "Attitudes and practices concerning the end of life: A comparison between physicians from the United States and from the Netherlands." *Archives of Internal Medicine* 160: 63–68.

Williams, Charlene, Matthew Rock, Eileen Considine, Seamus McCarron, Peter Gow, Roger Ladda, David McLain, Virginia Michels, William Murphy, Darwin Prockop, and Arupa Ganguly. 1995. "Three new point mutations in type II procollagen (COL2A1) and identification of a fourth family with the COL2A1 Arg519 → Cys based substitution using confirmation sensitive gel electrophoresis." *Human Molecular Genetics* 4: 309–312.

Woolgar, Steve. 1991. "Configuring the user: The case of usability trials." In *A Sociology of Monsters,* ed. J. Law. Erlbaum.

Wooster, Richard, Graham Bignell, Jonathan Lancaster, Sally Swift, Sheila Seal, Jonathan Mangion, Nadine Collins, Simon Gregory, Curtis Gumbs, Gos Micklem, Rita Barfoot, Rifat Hamoudi, Sandeep Patel, Catherine Rices, Patrick Biggs, Yasmin Hashim, Amanda Smith, Frances Connor, Adalgeir Arason, Julius Gudmundsson, David Ficenec, David Kelsell, Deborah , Patricia Ford Tonin, D. Timothy Bishop, Nigel Spurr, Bruce Ponder, Rosalind Eeles, Julian Peto, Peter Devilee, Cees Cornelisse, Henry Lynch, Steven Narod, Gilbert Lenoir, Valdgardur Egilsson, Rosa Bjork Barkadottir, Douglas Easton, David Bentley, P. Andrew Futreal, Alan Ashworth, and Michael Stratton. 1995. "Identification of the breast cancer susceptibility gene BRCA2." *Nature* 378: 762–763.

Wright, Susan. 1994, *Molecular Politics: Developing American and British Regulatory Policy for Genetic Engineering, 1972–1982*. University of Chicago Press.

Zola, Irving Kenneth. 1972. "Medicine as an institution of social control." *American Sociological Review* 20: 487–504.

An Engine, Not a Camera: How Financial Models Shape Markets
Donald MacKenzie

Building the Trident Network: A Study of the Enrollment of People, Knowledge, and Machines
Maggie Mort

How Users Matter: The Co-Construction of Users and Technology
Nelly Oudshoorn and Trevor Pinch, editors

Building Genetic Medicine: Breast Cancer, Technology, and the Comparative Politics of Health Care
Shobita Parthasarathy

Framing Production: Technology, Culture, and Change in the British Bicycle Industry
Paul Rosen

Coordinating Technology: Studies in the International Standardization of Telecommunications
Susanne K. Schmidt and Raymund Werle

Making Parents: The Ontological Choreography of Reproductive Technology
Charis Thompson

Everyday Engineering: An Ethnography of Design and Innovation
Dominique Vinck, editor

Index